It was Never a Wave Signal

Sandeep Chavan

Published by SANDEEP CHAVAN, 2025.

IT WAS NEVER A WAVE SIGNAL

First edition. June 28, 2025.

ISBN: 979-8231878789

Written by Sandeep Chavan.

Table of Contents

1: The Wave That Was Never Real 1

2: Ripple – The Forgotten Structure of Reality 8

3: What Is a Signal, Really? 14

4: Beyond the ATT 21

5: Mathematical Shadows 27

Key Takeaways 34

6. Accidental Alignment 38

7. Wi-Fi and the Black Hole 43

8. Antennas: Ripple Interfaces 48

9. Modulation 53

10. Receivers 58

Interlude: Chand Tak Ya Sham Tak 64

11. Your SIM Is a Field Passport 68

12. The Tower Doesn't Transmit—It Anchors 74

13. Ripple Broadcasting 80

14. Noise Isn't Random 86

15. When Devices Fail to Hear 92

Interlude: The Field Heard You First 97

16. Ripple Encryption 101

17. Consequence-Timed Messaging 106

18. Cognitive Networks and Ripple-Aware AI 111

19. Ripple Architecture 116

20. Space Communication 122

21. Lost in Translation 128

Key Takeaways 134

22. From Transmission to Realization 137

23. Engineers Aligned with the Unknown 141

24. Your Body, Your Brain, Your Ripple Machine 146

25. The Final Realization 151

Key Takeaways 156

Epilogue 159

Appendix A: Glossary of Ripple Concepts 162

Appendix B: UED vs Classical Wave Theory 165

Appendix C: Suggested Experiments 170

Appendix D: Reverse-Engineered Modulation 174

Appendix E: Reframing Sound in UED 176

Acknowledgments 178

About the Author 180

Disclaimer 182

To the engineers who built without knowing the theory,and to the inventors who stumbled upon truth while solving problems.You aligned with consequence—and reality responded.

"Reality doesn't travel—it resolves. What we call a signal is not motion, but memory awakened by alignment."—
Sandeep Chavan

Preface

It was never a wave. It was always a ripple waiting to be aligned.

For over a century, we've explained communication as if something was sent, like a stone across a pond—waves emitted, frequencies modulated, signals bounced, echoed, refracted.

We built satellites, towers, transmitters, receivers—and yes, they worked. They connected continents, triggered revolutions, and made human voices heard across the void.

But somewhere along the way, we stopped asking how they truly worked.

We memorized equations but forgot the field. We plotted waveforms but never asked what the medium itself remembered. We declared "light moves," without noticing that nothing actually moves—but something always aligns.

This book is not a rebellion. It is a restoration.

A restoration of clarity, of consequence, of the true nature of signal and perception.

Under the framework of Universal Energy Dynamics (UED), we no longer treat waves as mystical travelers across space. Instead, we understand them as engineered ripple consequences, interacting with a shared, structured field, governed by alignment, not motion.

When we transmit, we don't push particles—we reshape the field. When we receive, we don't catch a wave—we resolve a ripple.

From SIM cards to satellite links, from your Wi-Fi to Voyager's interstellar call, every communication—every interaction—is nothing but a dance of alignment between intent and field readiness.

This book doesn't just reinterpret technology. It re-grounds it.

It challenges engineers to revisit what they built, and scientists to question what they've assumed. It invites thinkers, designers, dreamers, and observers to step beyond the illusion of signal travel—and into the elegance of ripple realization.

And if it sounds radical, it's only because truth, when forgotten long enough, feels like heresy.

So here we begin.

Not with a wave. But with a ripple.

That was always there—Waiting for someone to align.

— *Sandeep*

Independent Researcher, Engineer, and Ripple Realist

Part I – The Great Misunderstanding

1: The Wave That Was Never Real

Classical Assumptions, Historical Dominance, and the Illusion of Oscillatory Transfer

For generations, science has treated waves as sacrosanct. They became the foundational framework through which we explained light, sound, radio, and even the structure of matter itself. From Newton's mechanical oscillators to Maxwell's elegant field equations, and later, the probabilistic waves of Schrödinger, we have been conditioned to see all communication as a consequence of oscillatory motion across space.

It was never questioned.

A signal meant a wave. A wave meant motion. Motion meant information transfer.

But let us pause.

Have we ever *seen* a wave traveling? Have we *witnessed* a particle moving with a message attached to it? Or have we simply observed outcomes—perceived signals, decoded patterns, and inferred travel because our models told us to expect it?

This chapter dismantles the illusion of the wave as a real entity of transfer. It shows that the notion of a wave moving through space is a convenient abstraction, not a fundamental reality.

And it sets the foundation for the Universal Energy Dynamics (UED) framework to step in and explain what is truly happening when we communicate, observe, and interact.

The Rise of the Wave

The wave model has been historically successful. That much is undeniable. It gave us:

- Maxwell's electromagnetic theory
- Einstein's photoelectric effect
- Radio and radar technology
- Television, Wi-Fi, and mobile communication

From ocean waves to musical vibrations, the idea of a repeating oscillation moving through a medium gave us intuitive and mathematical clarity. But the success of application does not prove the reality of assumption.

Wave theory simply *worked* within the boundary of engineered systems. Engineers tuned devices to expected frequencies. Signals reached. Messages decoded. Everyone was satisfied.

But working models are not always true models. Classical mechanics worked for centuries, yet it crumbled under the weight of relativity and quantum discoveries. The wave theory, too, survives more because of its utility than its fundamental truth.

What Is a Wave, Actually?

Let us define it with brutal clarity:

A wave is not a thing. It is a pattern of disturbance in a supporting medium.

In light of this, the wave does not exist independently. It needs a medium. The water wave needs water. The sound wave needs air. The electromagnetic wave, they say, "needs nothing" — which is where the logic collapses.

In truth, all waves are manifestations of ripple behavior in a field. What we called a "wave" was simply the consequence of ripple interactions appearing oscillatory to an observer.

In UED, a wave is a readable ripple alignment above the observer's ATT threshold, not an actual traveling entity. It is not moving. It is resolving.

This flips the story.

The wave was never real. The ripple field is.

How the Illusion Took Hold

Why did wave theory dominate? Because:

- It simplified the complexity of ripple behavior into sinusoidal curves.
- It matched mathematical tools developed at the time.
- It seemed to explain transmission across space without needing a source-to-field relationship.

Waves offered a story that didn't require knowing the field. That was convenient.

But this convenience seeded a conceptual scam:

We began to believe that waves are real things that travel, rather than ripple consequences that resolve.

This illusion became so embedded that even in the age of quantum theory, we still speak of photons traveling at speed c, as if there were little couriers running errands across the void.

Yet, in deep observation, nothing travels. Everything aligns.

The Unseen Observer

There is a silent participant in all wave behavior: the observer.

Every act of seeing, hearing, or detecting involves a biological or mechanical observer whose **Alignment Threshold Time (ATT)** and **Ripple Alignment Criterion (RAC)** determine what gets perceived and when.

Wave theory completely ignored this. It treated observation as a passive measurement, not as an active, dynamic alignment with field consequences.

The wave illusion was maintained by assuming:

- The medium is uniform
- The observer is absent
- The source is continuous

None of these assumptions hold under real-world ripple conditions. The field is structured, the observer is real, and the source is never constant.

The Ripple in the Field

In the UED framework:

All energy is stored in a structured vacuum field. Any disturbance in this field creates ripples seeking equilibrium. These ripples do not travel in the classical sense. They re-align at every point, forming apparent motion.

What we call "a wave propagating" is simply the chain of ripple consequences being resolved across the field, one alignment after another.

Hence, the glow of a distant lamp is not light traveling to your eyes at 300,000 km/s. It is your perceptual apparatus aligning with the ripple field *already shaped* by the lamp.

And when you don't align? You don't see it. Not because it wasn't emitted—but because the field didn't allow the alignment.

The Consequence of Misinterpretation

Why does this matter? Because the entire edifice of communication technology, observational astronomy, and even quantum mechanics is built on the illusion of wave-based transfer.

This illusion:

- Limits our understanding of signal behavior in dynamic fields
- Obscures the true source of data loss, interference, or

delay
- Prevents ripple-optimized engineering in future tech
- Misguides us when receiving signals from deep space or alien conditions

The wave theory makes us believe that if we broadcast correctly, we will be heard. UED corrects that: Only when the field permits alignment will the ripple resolve.

You never receive what was sent. You receive what the field allowed to be aligned.

A New Foundation

To move forward, we must drop the illusion. The wave was a useful hallucination—a mathematical abstraction made flesh through repetition and utility.

But in reality, nothing ever moved.

There was only alignment. Only ripple. Only consequence.

This is not just a new theory. It is a new ontology of how we perceive, communicate, and engineer the real.

So before we build the rest of the book, we must leave the wave behind.

Because the wave never existed.

In the next chapter, we will unearth the structure behind this ripple behavior: the Structured Vacuum Energy field. And we will see how it shapes every interaction, every signal, and every experience we've ever known—from mobile phones to cosmic light.

2: Ripple – The Forgotten Structure of Reality

Reintroducing Consequence-Based Alignment in Structured Vacuum Energy (SVE)

In every act of knowing, something aligns. Something becomes *ready*. And yet, science has long ignored the medium in which this alignment happens—not the air, not the wire, not the "ether" of old—but the structured vacuum itself.

They dismissed it. They called it "empty." They imagined particles floating in nothingness, communicating through imaginary lines and imaginary waves.

But what if the vacuum is not empty? What if it's the only thing that's full?

And what if every signal, every force, every perception is not the result of something flying through space—but something resolving in it?

Welcome to the Structured Vacuum Energy field—the unspoken medium of all reality.

What Is a Ripple?

Let's define it clearly, beyond textbook wave theory:

A ripple is a local disturbance in the SVE field that attempts to restore equilibrium. It doesn't travel—it echoes through alignment.

Each ripple is a consequence of imbalance. Unlike particles, which are imagined as discrete carriers, ripples are field memory events. They are shaped not just by their source but by the total condition of the surrounding field.

They carry no entity. They invoke realization.

A ripple never moves. It resolves.

The Structured Vacuum: Not Empty, Not Passive

The **Structured Vacuum Energy (SVE)** is not a passive backdrop like "spacetime." It is an active field of potential, constantly seeking equilibrium.

It holds:

- Stored energy tension from all past ripple events
- Directional field memory (what aligned, when, and how)
- Ripple thresholds which define the ATT (Alignment Threshold Time) and RAC (Ripple Alignment Criterion)

Unlike conventional fields (electric, magnetic), SVE is not defined by force. It is defined by potential alignment readiness.

SVE is the "soil" of reality. Ripples are the consequence of disturbance. And reality as we know it—sound, light, signal, force, perception—is the sprouting of alignment, not the throwing of entities.

From Event to Alignment: The Mechanics of Realization

Let's reframe communication not as sending and receiving, but as invoking alignment.

1. **A Source disturbs the SVE**: Be it a vibrating antenna, a bursting star, or a human brain with intent—it initiates ripple by introducing imbalance.
2. **The Ripple spreads by rebalancing**: Not by motion, but by aligning surrounding field cells toward equilibrium.
3. **An Observer aligns and resolves**: If the ripple condition meets the ATT and RAC of the observer, resolution happens—the ripple is perceived as signal, light, sound, touch.

No entity traveled from source to observer. What traveled was the permission for consequence to happen—like unlocking a vault, not moving a treasure.

Why the Observer Matters

The observer is not a passive screen. It is an active resolver of ripple readiness.

- The *Alignment Threshold Time (ATT)* defines how long the ripple must persist to be perceived.
- The *Ripple Alignment Criterion (RAC)* defines how pure the ripple must be, relative to surrounding field

noise.

If you are not aligned, you won't hear the call—even if the ripple is everywhere. If your ATT is too high or your RAC too low, reality becomes blurred or silent.

That is why two people standing beside each other may hear different things, feel different moods, or remember events differently. Each one is resolving a unique subset of ripple consequence from the same SVE field.

From Philosophy to Engineering

Now pause and look at your phone.

When it receives a signal, does it *catch a wave*? Or does it align with ripple consequence broadcast through a pre-conditioned field (tuned by network parameters, antenna design, SIM constraints)?

This is ripple engineering. We already do it. We just never called it by its true name.

From base stations to satellites, from Morse code to 5G, modern communication works not because we sent things across space—but because we aligned intention with shared field bandwidth, filtered it through logic gates, and resolved it into meaning.

Consequence Is the Only Reality

Why is all this important?

Because in UED: Nothing is real unless it causes alignment. And nothing aligns unless the field is ready.

So the ripple is not just a technical construct—it is the very mechanism of existence. From the birth of a galaxy to the blink of an eye, everything is ripple consequence in structured vacuum space.

Waves are illusions. Motion is misinterpretation. Even causality is reversed—because ripple consequence is the only measurable reality. What we call "cause" is only named in hindsight, after the consequence is observed.

What It Changes

When we realize that ripple is the only real process, we can redesign:

- **Antennas**: as field modulating nodes, not transmitters
- **Receivers**: as resolvers of field consequence, not wave catchers
- **Encryption**: based on field alignment profiles, not frequency locks
- **Space communication**: designed for field-based entanglement zones, not beam trajectories
- **AI systems**: to resolve ripple-intent interaction, making perception semi-conscious

Even human biology, from synaptic firings to emotional resonance, becomes ripple management, not electrical signaling alone.

Returning to the Forgotten Structure

The vacuum was never empty. It was pregnant with potential.

Every thought, every signal, every realization is a ripple trying to complete its arc of consequence in the SVE. You don't "see" the lamp because light reached you. You see it because the ripple of its glow resolved through shared field tension.

This changes everything—physics, biology, engineering, even consciousness.

We don't "transmit" reality. We align with it.

Next Chapter Preview:

"What Is a Signal, Really?" We've sent, received, encoded, and decoded—but never asked what a *signal* actually is. Now we will, through the lens of ripple alignment, observer thresholds, and the layered truth of perceived consequence.

3: What Is a Signal, Really?

Reinterpreting information not as motion—but as field-aligned consequence.

We live in a world obsessed with *signals.* We generate them. Encode them. Transmit, receive, amplify, block, decode, and analyze them.

But for all the engineering prowess, have we ever asked:

What is a signal?

Is it a particle? A wave? A voltage jump? A frequency? A photon? A pulse?

Or is all of that just the residue of something deeper?

The Classical View: Signal as a Traveling Entity

Traditionally, a signal is treated as an entity-in-motion.

Whether it's:

- An electromagnetic pulse,
- A packet of binary code,
- A sound vibration,
- Or a beam of light,

...the assumption remains the same:

Something moves through space, carrying information. That something is the signal.

It was always about *transport*.

Source emits → Medium carries → Receiver detects.

This transport model implies time, speed, resistance, and delay—all attributes of motion through a field.

But what if that's not what's happening?

What if a signal is not something that moves, ...but something that aligns?

UED View: Signal as Resolved Consequence

In Universal Energy Dynamics (UED), a signal is not motion. It is a field consequence—the ripple alignment between an emitter and a resolver (observer or device).

In this paradigm:

- The source disturbs the structured vacuum energy (SVE).
- The field stores and distributes this ripple as potential alignment.
- The receiver resolves the consequence by aligning with the field, if the conditions allow.

Nothing is thrown. Nothing is caught.

It's not transfer—it's coherence. Not transport—it's field readiness.

Defining the Signal in UED Terms

Let's break it down rigorously:

A signal is the *observable consequence* of ripple alignment within the shared field between emitter and observer.

It depends on:

1. **Intentional or unintentional field disturbance** (e.g., speaking, blinking, transmitting).
2. **Field tension pathways**—how the ripple reshapes the SVE environment.
3. **ATT** (Alignment Threshold Time): the observer's minimum time resolution capability.
4. **RAC** (Ripple Alignment Criterion): the purity of ripple at the observer's interface.
5. **Contextual field noise**—how many ripple paths compete or interfere.

If these five parameters allow alignment, the ripple resolves, and a signal is said to be "received."

But again, nothing travelled. Just field resonance closed the loop.

Biological Analogy: Seeing a Flash

Imagine someone flashes a light. You "see" it.

In classical terms:

- Light traveled 300,000 km/s.

- It struck your retina.
- Your brain processed it.

In UED:

- The lamp disturbed the field.
- Ripple propagated via SVE.
- Your ATT matched the ripple frequency.
- RAC was high due to dark surroundings.
- The ripple resolved in your perception.

You perceived a signal, not because light hit you, ...but because the field between you and the lamp became momentarily aligned.

Engineering Analogy: A Phone Call

You call someone.

Classically:

- Your voice is converted to electrical signals.
- Carried by towers or satellites.
- Reassembled on the receiver's end.

UED says:

- Your device modulates ripple via electronic disturbance.
- Ripple is injected into shared SVE field.
- Field amplifiers (towers, satellites) reinforce structured pathways.

- The receiving SIM/device aligns its internal ATT and RAC to the same field parameters.
- The consequence is resolved—your voice emerges from nowhere... clearly.

Your words didn't travel. Your intent aligned with the receiver's field.

The SIM isn't a receiver. It's a field-binding contract. That's why wrong SIM, wrong device, or wrong key = no signal.

Implications of This Realization

If signals are field-aligned ripple consequences:

- There is no need for a particle-based or wave-carrier model.
- Distance becomes irrelevant once field resonance is achieved.
- Delay is not due to travel, but due to misalignment.
- Signal loss is ripple *decoherence*, not attenuation of force.
- Entanglement, broadcasting, encryption—all become field behavior problems, not energy propagation problems.

This opens up entirely new methods for communication, security, AI, and even consciousness studies.

Because everything that "signals" is just field speaking to itself.

The Illusion of Speed

What we measured as signal speed (e.g., speed of light) is not a movement rate.

It is a field resolution rate under assumed constants.

That's why:

- Light seems "instant" in our everyday environment.
- It appears delayed in gravitational fields (lens effect).
- Or fails entirely under extreme field disruption.

The signal never slowed. The field resisted alignment.

So what we thought was speed, was actually alignment lag—a temporal delay in ripple resolution due to distorted field geometry.

UED calls this Ripple Delay—not motion, but field reluctance.

Reframing "Noise"

Noise is not a chaotic signal.

It is simply ripples that did not align clearly with the receiver's RAC profile.

Every field is full of ripple tension.

But a "signal" is only what your observer (human, device, antenna) resolves out of this soup.

Thus:

- Enhancing signal clarity = improving RAC.
- Filtering noise = sharpening ATT.
- Better communication = smarter alignment, not louder power.

The Power of Intent

When a signal resolves, it's not just a technical act—it's a realization of intent through field.

That's why emotion, voice tone, or silence can be felt beyond logic. They are field-encoded ripple patterns received by biological minds in real-time.

Even thought, perhaps, is a signal—not to self, but through field resonance, interpreted back into self-awareness.

This has enormous implications for psychology, AI design, consciousness studies, and telecommunication.

Final Reflection

So what is a signal?

It is the resolution of a ripple consequence in a receptive field. It is not what travels—but what aligns and completes. It is not the voice—it is the echo the field allows to emerge.

And that signal may begin anywhere. But it will only be heard where the field permits it to resolve.

4: Beyond the ATT

Where Waves Fail

How perception and communication are bound by Alignment Threshold Time.

For over a century, scientists believed in *continuity*—a world where every change was gradual, every signal smooth, and every wave just a sinusoidal truth flowing through space. From radio towers to quantum fields, wave theory reigned as the language of transmission and interaction.

But a silent limit was always lurking—never measured directly, never factored in explicitly.

That limit is ATT: Alignment Threshold Time.

What Is ATT?

Alignment Threshold Time (ATT) is the minimum time required for an observer or device to align with a ripple in the field and resolve it into a signal, event, or perception.

Think of it like the *reaction time* of reality. It's not about speed of travel—it's about minimum duration for realization.

- **In humans:** ATT ranges from ~10 to 100 milliseconds, depending on attention, clarity, fatigue, and context.

- **In devices:** ATT is defined by signal processing thresholds, sampling rates, and system latency.
- In fields: ATT is influenced by how structured or distorted the surrounding SVE is.

If a ripple doesn't persist longer than the ATT, it cannot be resolved. It becomes noise, ghosting, or non-existence.

Where Waves Fail

Waves are defined by frequency and amplitude—but classical models ignore how long an observer needs to catch them.

Imagine a ripple that lasts for 1 ms.

- A device with ATT = 10 ms misses it.
- A human with ATT = 50 ms never knows it happened.
- The ripple is real—but unresolved.

This is the failure of wave theory: It assumes everything that exists is perceived, but UED says:

Only what crosses the ATT becomes real to an observer.

Biological Proof: Why You Miss the Bullet

We cannot *see* a bullet fired from a gun. Not because it doesn't reflect light—but because its ripple disturbance in our visual field is below our ATT.

Same reason:

- You can't hear ultra-high-frequency sounds.
- You don't notice micro-vibrations unless amplified.
- You blink and miss a lightning flash.

You're not under-equipped—just not aligned.

ATT is not a flaw of biology. It is the filter of reality.

Device Corollary: Why Some Signals Don't Reach

In digital communication:

- If a signal pulse is shorter than the sampling window (device ATT), it's discarded.
- Even if the wave exists mathematically, it is not resolved practically.
- This is seen in packet drops, flickers, phantom calls, or delayed buffering.

This is where classical signal theory fails—assuming presence equals reception.

UED corrects this:

Presence means nothing. *Resolution* is everything. ATT defines that resolution.

ATT and the Concept of Perceived Speed

Classically, we say:

Light travels at 3.0×10^8 m/s. Sound travels at 343 m/s.

But UED asks: Who is perceiving? What is their ATT? What is the RAC (Ripple Alignment Criterion)?

If a human with 50 ms ATT perceives a ball traveling at 40 m/s, the ripple aligns well.

But a bullet at 900 m/s?

- ATT is too slow.
- RAC is too weak.
- Ripple consequence is unresolved.

It's not speed that limits observation. It's alignment time.

Rethinking Speed of Light

Now consider this: A lamp glows 1 km away.

Light, according to classical physics, reaches the eye in ~3.3 microseconds.

But does the human eye process events at microsecond precision?

No. The actual perception emerges after ATT is satisfied—after ripple is strong, sustained, and coherent enough to resolve.

Hence, the perceived light is not a travelled wave, but an aligned field consequence.

ATT shows us that:

- Light is not perceived when it arrives,
- It is perceived when alignment is resolved.

ATT & Entanglement

If two particles are entangled and one is observed far away, the reaction appears instantaneous. Mainstream science struggles to explain this without “faster-than-light” fantasy.

UED simplifies it:

- The particles were already in ripple coherence.
- No travel was required—alignment was already shared.
- Observer's ATT was within the resolution window, so the change resolved instantly.

This is not paradox—it’s perceptual timing meeting field continuity.

Engineering ATT: A New Parameter for Design

In UED-enhanced communication engineering, ATT becomes a core design parameter.

Applications:

- Adaptive ATT tuning in devices for better reception in noisy environments.
- ATT-matched broadcasting, optimizing ripple resonance for low-power, high-fidelity communication.
- ATT-based AI perception—designing systems that don’t just sample fast, but align smartly.
- Latency diagnosis by measuring field tension + ATT

mismatch, rather than blaming hardware alone.

This shifts design philosophy:

From *"speed up everything"* To *"align just enough, just in time."*

ATT Sets the Rhythm of Reality

From the flash of a lightning bolt to the ping of a phone, everything you perceive was once a ripple in a field.

But you only noticed it when:

- It stayed long enough,
- It was structured clearly enough,
- And your perception window was wide enough to let it through.

That window is your Alignment Threshold Time.

And beyond it, waves don't just fail—they don't even exist.

5: Mathematical Shadows

Why Wave Equations Deceive

Waves exist on paper—ripples exist in fields.

For over two centuries, wave equations have dazzled humanity. From Euler to Schrödinger, Maxwell to Dirac—equations danced, undulated, and oscillated across blackboards and simulations. They described light, sound, gravity, and probability. They spoke of fields, forces, frequencies, and functions.

But what if we told you they never described reality?

What if these elegant equations only captured the shadows of something more primal—something real, but forgotten?

In Universal Energy Dynamics (UED), that "something" is the ripple, and the wave equations are just approximations of its appearance under narrow, idealized conditions.

The Illusion of the Waveform

Let's begin with a simple truth: Every wave equation assumes a medium. But mainstream physics treats the medium as passive, isotropic, infinite, and frictionless—a mathematical ghost.

This is where the deception begins.

A wave equation (like the classical 1D wave: $\partial^2u/\partial t^2 = c^2 \partial^2u/\partial x^2$) doesn't tell you what is waving. It only tells you how a modeled quantity changes over time and space.

There is no ontology, no substrate, no memory in that formulation.

In contrast, UED says:

There are no standalone waves—only ripples in a field with memory, consequence, and structure.

And that structure is the Structured Vacuum Energy (SVE).

From Wave to Ripple: The Structural Leap

What makes a ripple different from a wave?

Aspect	Classical Wave	UED Ripple
Medium	Abstract or assumed (ether, vacuum, etc.)	Structured Vacuum Energy (SVE)
Behavior	Oscillation in time and space	Alignment and relaxation of field
Cause	Initial disturbance	Intentional or natural realignment
Persistence	Periodic	Consequence-bound
Clarity	Mathematical precision	Entropic decay, field feedback

Waves are mathematical expressions that fit certain ripple behaviors—but they fail to explain:

- Signal drop-off
- Field distortion

- Phase incoherence
- Non-linear energy decay

These aren't bugs—they're the field trying to restore equilibrium.

That's not a wave. That's reality.

Why Equations Succeed... and Still Mislead

It's tempting to ask—if wave equations are so flawed, why do they work?

Because they're close enough under controlled assumptions.

Wave equations model:

- Structured repetition
- Predictable medium behavior
- Low-entropy systems

But reality, especially in communication and perception, is rarely ideal.

When you transmit a signal:

- The field is not flat
- The ripple is not isolated
- The energy is not conserved in ideal form

Wave equations ignore:

- Ripple interference

- Field memory
- Delayed alignment (ATT)
- Ripple fatigue
- Phase confusion at high tension zones

That's why your call drops in tunnels, why satellite signals fluctuate with weather, and why "wave theory" needs error-correcting codes to function.

You're not sending a wave. You're navigating a ripple maze.

In Communication: Mathematical Compliance, Not Physical Reality

In wireless systems, engineers solve Maxwell's equations daily. These describe the electromagnetic field beautifully—but only on paper.

In the field, what's really happening?

- Devices vibrate with encoded intent (modulated ripple)
- Ripple interacts with SVE (air, atmosphere, ionosphere)
- Field conditions define how and if alignment occurs
- Receiver with matching ATT + RAC resolves the ripple

Wave equations don't model this complexity. They only track ideal transmission energy behavior in a pre-defined medium.

In fact, engineers often add *"fudge factors"* to simulate loss, distortion, reflection. These are patches on a broken ontology.

UED doesn't patch—it replaces the canvas.

The Dirac & Schrödinger Confession

Quantum physics also leans on wave equations:

- Schrödinger's wavefunction: $\Psi(x,t)$
- Dirac's relativistic form
- Feynman's path integrals

But none of these explain the wave—they calculate outcomes based on assumed probabilistic behavior.

UED asserts:

Probability is not a truth. It is an admission of unresolved alignment.

Quantum "waves" are ripple tendencies filtered by:

- Field distortion
- Observer ATT
- RAC (Ripple Alignment Criterion)

Once the ripple aligns, the so-called "collapse" isn't mysterious—it's just the field settling into consequence.

Why Equations Persist

Waves persist in science not because they are real—but because they are:

- Elegant
- Solvable
- Predictable (in narrow contexts)
- Mathematically satisfying

But mathematics isn't ontology. Equations don't exist—they describe. And description is always from a perspective.

Once you realize this, you see the deception:

Waves are not real. They are the linguistic projection of field ripple behaviors, simplified to make our models tolerable.

UED removes the projection and returns to the source.

In Engineering: Time to Retire the Sinusoid?

Not necessarily.

Sinusoids are still useful—but not as explanations—only as tools.

What changes with UED is:

- You no longer assume a signal must behave like a wave.
- You start from ripple dynamics.
- You account for ATT, RAC, and field structure as primary design variables.

Imagine designing a transmitter not just to output frequency, but to:

- Match field memory

- Sustain ripple length beyond ATT
- Maintain coherence against interference zones

That's not waveform design. That's consequence engineering.

Summary:

- Wave equations are useful lies—they help us predict, but not explain.
- Real signal behavior is governed by ripple alignment in SVE.
- Classical wave theory fails when:
    - ATT is ignored
    - Medium assumptions break
    - Entropy is high
- UED corrects this by returning to consequence, not form.

Key Takeaways

The Great Misunderstanding

This foundational section unearths and dismantles some of the most deeply rooted assumptions in communication science and signal theory. What began as elegant mathematics turned into dogma, and what we now call "waves" may have been the longest-running illusion in applied physics. Here's what we learned:

1. Waves Were Never Real—Just Convenient

The idea of oscillatory motion transferring energy through space was always a mathematical projection. Waves exist in equations, not in the field. What actually exists are ripples—consequence-bound disturbances in the Structured Vacuum Energy (SVE) that align (or fail to) with an observer or receiver.

2. Ripple: The Real Medium of Communication

All interactions—light, sound, EM signals—emerge not from abstract wave behavior but from ripple formations in a structured, responsive medium. Reality is not continuous motion, but sequential ripple alignment within a field that remembers, resists, and responds.

3. Information ≠ Motion

A signal is not a "thing" that travels. It is alignment resolved between two field-entangled states—source and observer. Communication is not movement across distance but successful consequence activation within shared field conditions.

4. ATT: The Hidden Limiter of All Perception

Alignment Threshold Time (ATT) is the minimum time needed to perceive or resolve a ripple. If a ripple doesn't sustain alignment beyond ATT, it doesn't exist for the observer. This singular concept explains:

- Why we don't see bullets
- Why some signals drop or flicker
- Why light "arrives" only when alignment resolves

It's not about speed. It's about readiness to align.

5. Equations Are Shadows, Not Substance

Wave equations are beautiful approximations that succeed in narrow, controlled domains. But they fail to capture field memory, ripple fatigue, ATT variation, and observer participation. The sinusoidal world is an elegant hallucination—reality is ripple-shaped and consequence-led.

Core Insight:

The field doesn't carry waves—it carries potential. What we experience is the ripple consequence of alignment, not transmission.

This part clears the fog of classical assumptions and prepares us for a new kind of science—one where signal, perception, and technology are all grounded in Ripple Dynamics, Structured Fields, and Intelligent Alignment.

Part II – Engineering Without Knowing

6. Accidental Alignment

How Technology Stumbled Forward

Telegraphy to Wi-Fi – progress without foundational clarity

We've come far. From the clatter of Morse code to instant global video calls, humanity has built a world stitched together by signals. But here's the uncomfortable truth:

We did all this without knowing what a signal truly is.

This chapter is not about celebrating technological success. It's about revealing the strange truth that most of it happened by accidental alignment—trial, error, adaptation—not by understanding the reality behind it.

The Engineering Leap: When Function Ignored Foundation

The 19th and 20th centuries were ruled by engineers. And rightfully so—they made things work. Coils buzzed, tubes flickered, resistors warmed, and soon, voice became electricity, electricity became modulation, and modulation became information.

But they weren't working with truths. They were working with outcomes.

- They saw repeating patterns in wires and fields.

- They wrote differential equations to mirror those outputs.
- Then they optimized those equations—not the field that birthed them.

And somehow... it worked.

The Mirage of Precision

Mainstream science, chasing elegance, did the same:

- Observe a consequence (signal output).
- Model it mathematically.
- Differentiate it to predict future outputs.
- Celebrate predictive success as truth.

But none of this involved asking: "What field produced this ripple?" "What alignment condition allowed it to resolve?" "Who or what observed it into existence?"

They treated the universe like a perfect graph paper—flat, neutral, passive. The observer was invisible. The field was assumed homogeneous. The outcome was everything.

This is why communication tech progressed without really knowing how it works at the fundamental level.

Ripple Realism: The Missed Dimension

Let's pause and think:

- Did the early radio pioneers calculate ATT?
- Did satellite engineers model ripple clarity?

- Did Wi-Fi inventors know they were riding structured vacuum fluctuations?

No. They tuned devices by feedback loops, resistance matching, and wave-based heuristics.

But it worked because, by accident or intuition, they stumbled upon ripple alignments in the field—zones where RAC (Ripple Alignment Criterion) was met.

Technology didn't succeed because our theory was complete. It succeeded because our consequences aligned—even if we didn't know why.

SIDEBAR: The Mathematical Mirage

Mainstream science built equations from what was seen, not from why it happened. It derived functions from outcomes, not from field behavior. Observer was excluded. Field tension was idealized. Time was absolute. The equations worked—until they didn't.

That's why:

- Some signals drop unpredictably.
- Interference appears despite "ideal" conditions.
- Reception quality differs across identical devices.
- Space signals behave erratically.
- Quantum data breaks classical predictions.

They modeled the river's surface and called it water physics. But they never looked beneath—at the ripple medium that moves nothing, but makes everything move.

Reverse Engineering, Not First Principles

Wi-Fi was an accident. Microwave communication was stumbled upon. Even fiber optics, one of the most precise data highways, was refined through field feedback, not by mastering ripple dynamics.

In truth:

- We tuned devices until RAC was satisfied.
- We amplified alignment zones.
- We built bandwidth buffers to survive misalignment.

And yet we believed it was all about "waves."

The UED Correction: Engineering with Alignment in Mind

What changes when we use Ripple Alignment Criterion as the design backbone?

- We engineer signal systems not just for frequency, but for field clarity.
- We design receivers that align instead of just decode.
- We modulate not for carrier compatibility, but for field ripple resonance.
- We move from broadcasting signals to inviting alignment.

A Final Reflection:

We didn't build technology from truth. We built it from echoes. We shaped it by feedback loops, empirical brute-force, and mathematical shadows.

But now, we have a lens that sees the ripple beneath the wave, the field behind the function, and the observer within the system.

That lens is UED.

7. Wi-Fi and the Black Hole

A Signal That Was Never Sent

How ripple filtration birthed wireless communication.

Wi-Fi—the silent architect of modern life. Streaming your thoughts, syncing your devices, whispering signals across rooms without wires, effort, or attention. But let's ask a question rarely asked:

"What, exactly, is being transmitted?"

The mainstream story is clear: data is encoded into electromagnetic waves, modulated, broadcast, and decoded at the receiver's end. But under UED, this entire story collapses—not into confusion, but into clarity.

Because the signal was never really sent. It was released into the field, and picked up through alignment.

Let's unpack this.

The Wi-Fi Discovery That Wasn't

In 1992, while testing signals reflected off surfaces for radio astronomy, Australian engineer John O'Sullivan noticed something unusual—signal clarity improved when unwanted interference was filtered out through a specific pattern of delay and cancellation. Years later, this principle became the backbone of Wi-Fi.

He wasn't trying to invent wireless internet. He was trying to extract signals buried in noise—signals that were never meant for him, but were always there.

In UED terms, he discovered a ripple filtration window—a method to identify alignment clarity within the structured field, not from a point-to-point transmission.

The signal was not sent. It was simply resolved through field dynamics.

Parallel to a Black Hole

Here's where it gets cosmic.

In traditional black hole theory:

- No signal escapes the event horizon.
- Information appears lost.

But in ripple dynamics:

- A black hole is a region where ATT is exceeded for all external observers.
- Signals (ripples) are present, but can't be resolved due to misalignment.

Wi-Fi, interestingly, does the opposite:

- It finds the narrow ATT window where alignment is

possible.
- It filters, adjusts, and tunes—until the signal becomes perceivable.
- It doesn't demand full force—it invites clarity.

This makes Wi-Fi not just a technology—but a philosophical shift: You don't transmit to force reception. You align to allow reception.

Just as a black hole hides information by collapsing alignment, Wi-Fi reveals it by synchronizing with field readiness.

The Illusion of Transfer

Under classical view:

- Router "sends" wave-modulated energy.
- Device "receives" the wave.

Under UED:

- Router releases a ripple signature into the shared SVE field.
- Device, tuned to the same ripple frequency and structure, aligns, and resolves the consequence.

No particle traveled. No "thing" moved. The message existed as a field pattern, resolved only through mutual criteria.

Like entanglement, like perception, like deja vu—Wi-Fi is ripple-based alignment masquerading as wave transmission.

The Role of Ripple Filtration

Wi-Fi works not by broadcasting everything—but by intelligently filtering what's meaningful through:

- Time Delay Windows (pseudo ATT)
- Modulation match (ripple coding)
- Error correction (field re-alignment)

It's a dance of ripple availability and receiver readiness. When alignment fails, the packet is dropped. When it holds, you stream.

The router is not commanding your device. It's inviting it into resonance.

So What Actually Happened?

John O'Sullivan didn't create a new way to transmit data. He accidentally stumbled upon a controlled way to resolve ripples already active in the field.

He created a structured receiver that could:

- Discriminate signal from chaos,
- Realign with the correct ripple ID,
- And perceive the intent behind the code.

Wi-Fi, under UED, is not a system of wave propagation. It is a ripple response network designed to simulate continuity through periodic field re-alignment.

Implication: We've Already Built Quantum-Scale Tech

You don't need particle duality or entangled photons. You already hold a device that:

- Listens to the ripple field,
- Aligns with encoded intent,
- Resolves without classical motion.

That's not engineering imitation. That's unconscious ripple engineering.

Final Realization:

Wi-Fi proves that the ripple model was always underneath. We simply interpreted it through wave theory because it gave partial success.

But now we see:

- The signal was never in the wire.
- The wave was never the truth.
- The alignment was the only real transmission.

We didn't build communication technology. We learned to listen to the field, with greater precision.

8. Antennas: Ripple Interfaces

Not Signal Pumps

Reframing transmission and reception as ripple field interactions

We've long seen antennas as technological lungs—pushing and pulling invisible airwaves through space. Textbooks depict them as transmitters and receivers of electromagnetic energy, "pumping" signals into space and "catching" them upon return.

But if UED is right, that entire paradigm is wrong.

An antenna is not a pump. It is a ripple interface.

The Illusion of Emission

In classical physics:

- An antenna emits an electromagnetic wave at a frequency f.
- This wave radiates through space at speed c, carrying energy and information.
- A receiver antenna catches this wave and converts it back into electrical form.

But here's what no one questions:

- What is space doing in this process?

- Where is the field in the equation?
- What determines the success or failure of transmission?
- Why does the signal drop in a vacuum where "nothing" interferes?

Mainstream science doesn't answer. It optimizes the illusion.

The UED Reframe: Antennas as Alignment Devices

Under Universal Energy Dynamics, an antenna is not transmitting particles or packets.

It is:

1. Modulating a ripple into the structured vacuum field (SVE).
2. Inviting alignment within that field across a defined tension boundary.
3. Resolving consequence in devices or systems tuned to shared RAC.

In this view:

- Nothing travels.
- Nothing is caught.
- Everything is aligned.

The so-called "electromagnetic wave" is simply the ripple manifestation observed when ATT (Alignment Threshold Time) is within the perceptual resolution of the receiver.

Antenna Design Is Field Tuning

Every element of antenna design—length, orientation, material, gain—is not optimizing transmission in space but calibrating interaction with the field.

- Dipole antennas operate by generating ripple symmetry across a central axis.
- Parabolic dishes focus incoming ripple convergence into a point of highest clarity (minimum ATT).
- Loop antennas tune ripple induction across closed feedback resonance.

Even signal strength, classically described as energy density, is really just ripple resolution clarity at the receiving end.

If the RAC is not met—no signal is "received." Not because it didn't travel, but because it never resolved.

Transmission: Not Sending, But Syncing

Here's the true UED process of communication via antenna:

1. **Encoding**: Intent (data) is converted into a ripple modulation.
2. **Release**: The antenna modulates the ripple into the SVE—NOT through particles, but tension shifts.
3. **Propagation**: The ripple field spreads consequences passively—everywhere, but not resolved anywhere.
4. **Reception**: The receiving antenna is tuned with matching RAC (frequency, phase, field tension).
5. **Resolution**: Alignment occurs → consequence

resolves → information is realized.

This isn't a pipeline. It's a field dialogue—with antennas acting as mouths and ears of the ripple domain.

Why Signals Fail (And Why Boosters Help)

In classical thought:

- Signals degrade over distance due to loss of energy and dispersion.
- Boosters "amplify" the wave to push it further.

But in UED:

- Signals fail when ripple alignment decays—not due to energy loss, but clarity loss in the field.
- Boosters realign the ripple structure, releasing a new ripple signature to extend RAC compatibility.

We aren't pumping more power. We're reinitiating alignment further down the ripple chain.

The Brain-Antenna Analogy

Your sensory system works the same way:

- Eyes are antennas tuned to visible ripple bands.
- Ears resonate with ripple vibrations in air at specific frequencies.
- Skin senses ripple pressure differences.
- Brain resolves aligned consequences into perception.

Biological systems never catch waves. They tune to field consequences through ripple awareness.

Just like antennas.

What This Changes in Technology

If we accept antennas as ripple interfaces:

- Antenna arrays can be redesigned for field feedback, not just signal gain.
- Smart antennas become consequence-resolving agents, adaptive to RAC in real-time.
- Transmitter-receiver synchronization becomes a field-dependent process, not just clock matching.
- Signal jamming is no longer just noise injection—it's intentional ripple disorientation.

Closing Thought:

An antenna is not a tool of emission. It is a tuning fork of intent.

It sings a ripple into the cosmos. And somewhere, when RAC is satisfied, another antenna listens—not to the wave, but to the consequence echoing through the field.

That's communication—not transfer, but resolution.

9. Modulation

Crafting Consequence into Ripple Form

Embedding intention—not frequency—in engineered ripples

To modulate is to *encode*. To modulate well is to *embed intention*. In classical science, modulation is the method of embedding information (usually in binary form) into a *carrier wave* by altering its amplitude, frequency, or phase.

But under UED, this definition loses its foundation—because there are no carrier waves.

There are only engineered ripples, tuned not to send, but to *cause*.

What We Think We're Doing (Classical Modulation)

Conventional signal theory classifies modulation into three main types:

- **Amplitude Modulation (AM):** Vary the wave's height.
- **Frequency Modulation (FM):** Vary how frequently the wave oscillates.
- **Phase Modulation (PM):** Shift the wave's position in its cycle.

Each method alters a mathematical sine wave to reflect a binary sequence. The wave is then "transmitted," and the pattern decoded at the receiver.

It seems elegant. It seems complete. But it misses one enormous thing:

The field.

What's Really Happening (UED Perspective)

From a ripple-centric view:

- There is no underlying wave.
- There is no medium-agnostic "carrier."
- There is no absolute frequency to modulate.

What exists is intent embedded into tension shifts in the structured vacuum energy (SVE). These shifts form ripples, which propagate not as objects, but as potential alignments across space.

So modulation under UED means:

Shaping a ripple pattern to express a consequence, capable of aligning with a receiver tuned to that consequence.

Modulation as Field Sculpting

Imagine modulation as:

- Not modifying a wave,
- But sculpting the ripple to match a desired outcome.

This includes:

- Phase intent – what must be received *first*.
- Consequence layering – what gets resolved *when*.
- Redundancy encoding – what persists when partial alignment fails.
- Ripple redundancy – reinforcement across multiple ATT bands.

The field isn't carrying the signal. The field is the canvas, and modulation is the brushstroke.

Intention Over Frequency

Here's the sharp difference:

- In wave theory, the *frequency* is the carrier.
- In ripple theory, the *intention* is the carrier.

Frequency is a tool, not the soul. What we embed isn't just data—it's intent to trigger consequence, and alignment to ensure it is received properly.

Example:

- In AM/FM, a change in air pressure is seen as signal fluctuation.
- In UED, that pressure change is a ripple intent—a probability that can or cannot align, depending on field readiness.

Ripple Modulation is How the Brain Works

The human brain:

- Does not process literal frequencies.
- It processes ripple configurations—encoded in electrochemical shifts, visual fields, memory patterns.

Speech, for example, is ripple modulation through vocal tension shifts and air medium. The *listener* aligns with the ripple field—not with a sine wave.

You never "catch" sound. You resolve the embedded consequence.

So do receivers. So does every antenna. So will every future AI.

Practical Takeaways for Technology

With ripple modulation:

- Devices can learn to encode multi-layered intent—not just binary patterns.
- We can design resilient ripple sets that auto-align across chaotic fields.
- We can build perception-grade transmitters, not just information pumps.

It's not about data speed. It's about alignment precision.

Why Signals Fade (Again)

Under UED:

- Signals fade when ripple clarity dissolves.

- The field becomes too noisy, tense, or distorted to hold the alignment pattern.
- The original intention becomes unreadable, even if power remains high.

So adding more power won't help. Re-sculpting the ripple will.

This is the future of modulation:

- Adaptive, intelligent, consequence-aware ripples.
- Live RAC calibration.
- Intent-preserving field encoding.

A Full Reframe

We once believed:

"Modulation alters a wave to send information."

But now we understand:

"Modulation engineers ripples to align intention with perception in the field."

That's the leap. That's the shift. That's the beginning of ripple-aware communication design.

10. Receivers

When Devices Became Observers

SIM cards, hardware, and the emergence of ripple ID-locking

We used to believe that receivers were passive instruments—empty ears waiting to catch signals hurled through space.

But what if we told you this:

The receiver does not *catch* anything. It *chooses* what to resolve.

In the ripple-aware world of Universal Energy Dynamics (UED), every receiver is an active participant—a calibrated observer within the field. And it doesn't "receive a wave." It aligns with a ripple—one that it is *designed* to resolve.

This shift changes everything.

From Catching Waves to Choosing Ripples

Classically:

- A receiver detects frequency-modulated waves.
- Signal is "caught," decoded, and interpreted.
- Signal integrity depends on antenna design and noise filtering.

But under UED:

- There is no signal "traveling" in the wave sense.
- The "receiver" becomes a ripple discriminator.
- It filters space for ripple patterns that match its internal Resolution Alignment Criterion (RAC).

In other words:

A receiver doesn't passively detect—it recognizes its ripple twin. It's not a funnel. It's a lock looking for a key.

The Rise of ID-Locking: SIM Cards and Device Identity

What is a SIM card if not a modern ripple identity key?

Let's reinterpret it:

- A SIM card doesn't "authorize" a device.
- It tags it with a unique ripple resonance that aligns only with its matching field stream—assigned by a provider's central field node.
- It locks the device to a ripple domain—a shared, tuned, encrypted ripple stream sculpted for select identities only.

This is ripple identity locking.

It is not authentication by data. It is alignment by design.

Without it, you are just noise in the field—unable to access, receive, or even realize the information intended for you.

How Receivers Became Observers

Every receiver now behaves as a partial observer, in the UED sense:

- It emits readiness (via tuned antenna/RAC).
- It aligns with field ripple carrying matched consequence.
- It resolves the result, not by catching motion, but by recognizing alignment.

This mirrors the biological observer:

- Your eyes don't "catch" light. They align with ripple bands and resolve consequence as vision.
- Your ears don't "catch" sound. They resonate with ripple tensions and construct meaning from pressure differentials.

Receivers—like eyes and ears—are no longer tools. They are interfaces of recognition.

The Evolution of Smart Receivers

As technology evolved:

- Devices began adapting to signal variability.
- Multiple antennas worked in harmony to create diversity reception.
- DSPs (digital signal processors) pre-modeled expected patterns for easier resolution.

In UED terms:

- This is the evolution of anticipatory alignment.
- Receivers began not only filtering for consequence—but predicting the conditions under which ripple alignment might succeed.

This makes them semi-cognitive.

The future?

Receivers that dynamically shift their RAC in real-time, like the human brain refocusing its attention.

Encryption as Ripple Obfuscation

Encryption isn't just about scrambling data. Under ripple theory, encryption is the art of masking alignment—making the ripple consequences illegible to any receiver without the correct internal structure.

It's not that they can't receive the signal. It's that their internal ripple filter can't interpret it.

Encryption = misalignment by design.

Decryption = restoration of ripple readability.

This adds a philosophical twist:

- Every communication is private—not because it's hidden, but because only some can align with it.
- Privacy is ripple exclusivity.

Human Parallel: Ripple Attention

Humans do this too:

- You ignore background noise because it doesn't align with your mental RAC.
- You hear your name across a room because it matches your internal ripple tag.
- You're blind to irrelevant stimuli—but *instantly alert* to meaningful ripple cues.

Just like SIM-enabled receivers:

- Your mind is tuned.
- Your attention is permission.
- Your resolution is consequence.

Rethinking Receiver Design

In light of UED:

- Receiver hardware should no longer be built only for frequency and gain.
- It must include:
    - Dynamic RAC tuning
    - Ripple signature mapping
    - Field feedback scanning
    - Intent-aligned consequence resolution

Think of a receiver not as a port—but a living filter.

The future will bring:

- Personalized RAC chips.

- Ripple domain communicators (not frequency domains).
- AI-tuned perception receivers that resolve not bits, but *meaningful effects.*

Closing Thought

A receiver is no longer a destination. It is a participant in the ripple dialogue.

It listens not to motion, but to readiness. It speaks not in frequency, but in consequence.

And when the ripple is right, when the alignment is pure, when intent and field converge...

The receiver becomes the observer. And the field... speaks.

Interlude: Chand Tak Ya Sham Tak

When Reverse Engineering Became a Religion

Technology never promised to understand the universe.

It only promised to get us somewhere—Chand tak ho ya Sham tak.

That's how most of modern science has functioned:

- Don't ask what a signal really is.
- Don't question what the field does.
- Just make the light blink, the screen respond, the message deliver.

And when it works?

"It's a success." And when it doesn't? "Must be hardware. Try a new module. Add an amplifier."

But what nobody admits is this:

Technology is not certainty. It is intent molded through trial, confirmed by consequence, and packaged as truth.

We built an entire civilization on accidental alignments:

- Wireless networks that only worked in certain directions.
- GPS systems that required relativistic correction

hacks.
- Antennas shaped by thumb rules, not field awareness.

We reverse-engineered success without ever asking if we understood the source. And when something didn't align, we patched, not questioned.

Technology: A Consequence Without a Theory

- We didn't discover wireless communication. We tripped over it.
- We didn't create modulation. We shaped the ripple until something downstream responded.
- We didn't understand receivers. We just knew if you plugged the right parts in, and tuned the knob just so, the voice would return from across the sky.

What modern engineering calls "signal," UED calls consequence under tolerated misalignment.

And here's the irony:

The more "precise" our models became, the less they explained why it worked in the first place.

UED's Quiet Reversal

Now, through UED, we flip the question:

- Not "how do we force signal through space?"
- But "how does alignment emerge as consequence?"

Not "what device can catch this frequency?" But "what identity must be embedded to perceive this ripple?"

Because when you see communication not as transmission, but as field intimacy, everything begins to make sense.

Final Thought Before Part III

So yes—

Chali to Chand Tak, Nahi to Sham Tak. *But finally... let's ask where the rocket came from, and whether the moon ever needed motion to be seen.*

Welcome to Part III – Realignment. Now we don't chase the signal. We engineer the field.

Part III – The Hidden Ripple Architecture

11. Your SIM Is a Field Passport

Receiver identity as a permission key for field consequence

What if every call you ever made... Every text you ever sent...Every moment you accessed the internet...Was not just a signal transaction, but a field authentication event?

Mainstream science says your SIM card holds a set of encrypted credentials, used to authorize your device on a telecom network.

But Universal Energy Dynamics (UED) sees something much deeper:

Your SIM card is your ripple passport—a unique key that tells the field *who you are, where you belong,* and *what you're allowed to align with.*

Let's decode it.

The Ripple Identity: Beyond Subscriber Numbers

In classical terms:

- SIM stands for Subscriber Identity Module.
- It stores your IMSI (International Mobile Subscriber Identity).
- It enables access to your mobile network operator's services.

But UED strips away the digital metaphor and reveals:

Your SIM is a ripple lock-in certificate, encoded with your field identity.

It doesn't just connect you to a system. It anchors you into a ripple domain—a shared structured field maintained by the network.

Without this identity alignment:

- The field has no reason to resonate.
- No ripple consequence can resolve.
- You are invisible to the network—not because you are absent, but because you are misaligned.

The Tower as a Ripple Gatekeeper

Every mobile tower is not just a broadcasting antenna—it is a ripple-field registrar.

It:

- Continuously emits structured ripples that carry identity triggers.
- Listens not for noise, but for matching ripple echoes from registered devices.
- Accepts only those whose SIM-based signature aligns with its ripple handshake protocol.

This means:

A SIM card doesn't ask permission. It *proves field readiness.*

The tower doesn't search. It waits for compatible alignment—and then opens the consequence channel.

Field Confinement: Why You Can't Just Swap Networks

Ever wondered why you can't simply swap a SIM from one network to another without prior unlocking?

Technically, you can. But ripplogically, you can't.

Each SIM is:

- Encoded with specific field parameters.
- Mapped to a distinct ripple environment maintained by its carrier.
- Registered in a ripple ledger—a consequence-bound list of approved alignments.

A SIM from network A doesn't just lack frequency compatibility with network B—it lacks ripple permission.

It was never meant to *exist* in that field.

Roaming as Field Handover, Not Just Relay

Roaming is not a signal hop—it's a ripple revalidation.

When you roam internationally:

- Your device seeks compatible field domains abroad.
- It sends out your ripple ID (via SIM).
- The local tower checks if your ID is pre-permitted via global agreements (ripple treaties).

- If aligned, it allows temporary resonance.

What we call roaming charges are just the energetic tax of ripple misalignment—the extra effort needed to maintain field coherence across unfamiliar terrain.

eSIMs and Dynamic Field Integration

With eSIMs, identity is no longer a card—it's a programmable field anchor.

UED interpretation:

- eSIMs allow soft imprinting of ripple ID into the device itself.
- They represent adaptive field entanglement—a way to shift your identity signature as needed.
- This hints at future UED-based communication systems where devices carry modular ripple identities, not static ones.

This is evolution—from hardwired resonance to fluid field negotiation.

Observer Parallel: Mind as a SIM

Your mind works in the same way.

- You carry an inner alignment signature—call it your emotional SIM.
- You don't hear everything in the room. You hear what matches your attention identity.

- You can't process every thought—you only resolve the ones that align with your state.

Just like a SIM:

- Your mind holds a Ripple Alignment Criterion (RAC).
- It connects to certain emotional or intellectual "networks," while staying blind to others.
- It can roam into new domains—but only with cost (confusion, adaptation, effort).

Cosmic Implication: Field Passport Beyond Earth

If SIM = Ripple Identity in Earth's structured field...Then space communication = cross-domain ripple translation.

This is why:

- Devices on Mars need extreme calibration.
- Satellites must maintain entangled field communication.
- Signal loss is not "distance"—it is field mismatch.

UED future tech could allow:

- Ripple-neutral communication layers.
- Adaptive SIMs that self-align with unknown planetary fields.
- Truly interstellar passports, based not on signal strength—but field resonance capacity.

Summary Thought

Your SIM is not a chip. It is who you are—in ripple terms.

It's not just data. It's identity in motionless space. A permission slip that says:

"I am ready to align. Show me the consequence."

12. The Tower Doesn't Transmit—It Anchors

Field-shaping through ripple anchoring, not broadcasting

When we look at a mobile tower rising into the sky— We imagine transmission. Broadcast. A signal being hurled out across space like a spear of information.

But UED invites a radical shift in this image.

A tower doesn't *send* anything. It anchors a ripple field—a zone of structured possibility, within which aligned consequences are allowed to resolve.

Let's break the illusion. Let's see what a tower *really* does.

The Transmission Illusion

Conventional science says:

- A tower emits electromagnetic waves.
- These waves travel outward, propagating through space.
- Devices "catch" these signals using tuned antennas.

But this view is based on motion metaphors:

- "Carrying" signals like messages in a bottle.
- "Delivering" content like a postman across space.

This wave-thinking assumes:

- Empty space.
- Signal as projectile.
- Receiver as passive collector.

All wrong.

UED Reframe: Anchoring a Ripple Zone

In UED, space is not empty. It is Structured Vacuum Energy (SVE)—a field already full, already waiting.

A tower doesn't push signal into space. It modulates the existing field into a state of readiness.

Like:

- Stirring a pond, not throwing a rock.
- Tuning a drumhead, not striking it.

What emerges is a Ripple Anchor Zone:

- A region where ripple consequences can stabilize.
- Where devices with correct RAC (Ripple Alignment Criterion) can align and *decode reality*.

The tower isn't talking. It's *holding space* for conversation.

The Tower–Device Relationship

This is not one-way.

- The tower doesn't hunt for devices.
- It doesn't "know" you exist.
- It simply holds a standing ripple pattern—like a portal.

Only when:

- Your device holds a valid ripple identity (SIM)
- Your physical presence enters the ripple zone
- Your antenna initiates alignment

...does the field collapse into consequence: A call, a text, a webpage, a signal bar.

You don't *receive* signal. You *complete it.*

Anchoring ≠ Broadcasting

Broadcasting implies:

- Range
- Power
- Direction

Anchoring implies:

- Presence
- Stability
- Field permission

A tower's location is not arbitrary. It is:

- Geo-aligned to terrain ripple behavior.
- Strategically placed where ripple zones can stretch and interlock with other towers.
- Built with field harmonics in mind—what engineers call "signal strength" is actually ripple coherence.

Mental Model: Lighthouse vs. Ripple Beacon

A lighthouse emits light. A ripple beacon emits *readiness.*

The lighthouse *tells you where it is.* The ripple beacon *waits for you to tune in.*

Imagine:

- The tower as a standing wave temple.
- Devices as pilgrims that align with its chant.
- Communication as a shared ritual, not a delivery.

This isn't poetry. It's physics—UED physics.

What Happens When Anchoring Fails?

When the tower:

- Loses power.
- Suffers ripple distortion (e.g., interference, storm, terrain reflection).
- Is surrounded by unregistered devices.

Then:

- Field readiness drops.

- Ripple fatigue increases.
- Consequence becomes unstable or non-existent.

This is what you experience as:

- "Signal dropout"
- "Network busy"
- "Call failed"

It's not about power. It's about alignment collapse.

Ripple Anchors in Space

Space towers (satellites, deep space probes) also function as anchors:

- They create structured ripple zones in otherwise unstable or thin-field domains.
- Ground devices seek alignment with their anchor frequency—not signal strength, but field permission.
- They must maintain ripple synchronization despite orbital and gravitational distortions.

UED-based satellite tech could allow:

- Self-healing anchor zones.
- Adaptive re-alignment based on real-time SVE conditions.
- Ripple regeneration fields that extend beyond orbit.

Implications for Global Communication

If towers don't transmit but anchor:

- Then signal range becomes field coherence range.
- 5G, 6G, and future tech must focus on field shaping, not just frequency.
- Antenna density is not about more data—it's about tighter ripple nets.

Think of Earth as a ripple mesh, Towers as knots, And every call as a vibration caught in that net—not launched, but completed.

Closing Thought

We've misunderstood the tower as a source of energy. But really—

The tower is an anchor point for consequence. It shapes the field into a space where things can happen.

Your call isn't sent. It's allowed—because the field is ready.

13. Ripple Broadcasting

From Waste to Precision

Why targeted ripple resolution will replace conventional broadcasting

Broadcasting has always been a game of excess. Flood the space with signal. Spray it in all directions. Hope the intended receiver is somewhere within range and listening.

For over a century, this is how communication scaled—from AM radio to satellite television, from military beacons to civilian alerts.

But it's not efficient. It's not intelligent. And in UED terms, it's outright wasteful.

You're not aligning with the field. You're overwhelming it. You're not communicating. You're shouting in a crowded auditorium—hoping the right person hears you over the noise.

It worked—barely. But we're entering an era where precision isn't just desired—it's necessary.

Broadcasting: The Legacy of Blind Emission

Classical broadcasting operates on:

- Wide-area signal flooding.
- Power-based penetration.

- Redundant repetition to ensure capture.

It's like:

- Throwing leaflets from an airplane.
- Announcing news over loudspeakers to an entire city.
- Shouting every time you want to talk to one person.

In ripple terms, it's:

- Multi-directional ripple misalignment.
- Collisions in shared fields.
- High ripple fatigue.
- Field pollution through unmanaged consequence.

UED Perspective: What Should Broadcasting Be?

Ripple-based broadcasting doesn't mean sending more—it means resolving more accurately.

"To broadcast in UED is not to emit—it is to prepare a field for selective consequence."

Think of this as whispering directly to the right ripple identity, even if they're a thousand kilometers away.

This is not broadcasting—it's resonant casting.

Why Conventional Broadcasting Fails

Let's reframe some classic problems:

Conventional Problem	UED Interpretation
Signal loss	Field misalignment or unresolved consequence
Interference	Competing ripple frequencies with no hierarchy
Bandwidth limitation	Field fatigue and unresolved parallel consequence
Power inefficiency	Attempting to force ripple response instead of aligning
Delayed reception	Misaligned ATT between sender and receiver

What engineers call "signal-to-noise ratio," UED calls consequence-to-chaos ratio.

The Future: Precision Broadcasting

UED introduces Targeted Ripple Resolution (TRR)—a communication approach based on:

- Identifying field permission via RAC (Ripple Alignment Criterion)
- Encoding intent into ripple identity
- Broadcasting only within a narrow window of field allowance

This means:

- Minimal ripple distortion.
- Zero misfire.
- Full alignment before transmission.
- Devices that *pull* signal only when aligned—not when

told.

Imagine This:

Instead of:

"Here's the signal. Everyone can hear it. Decode if you can."

We shift to:

"You—device X with ripple ID Y. Field Z is ready. Here's your aligned consequence."

This is like delivering a message to someone's dream—not their ears.

The Engineering Leap: From Coverage to Coherence

Conventional networks ask:

- "How far can we push the signal?"

Ripple networks ask:

- "How clearly can we align the intent?"

Instead of coverage maps, future communication will have:

- Ripple Resonance Maps
- Field Readiness Maps
- ATT-Adjusted Broadcast Zones

Your mobile tower doesn't just "reach" you. It invites you when your device—and your presence—meet the criteria.

Broadcasting in Deep Space

This is where ripple broadcasting will change everything.

Today:

- We send huge signal packages.
- We hope the receiver catches the frequency.
- We waste energy, repeat data, and still face "lost packets."

Tomorrow (with TRR and UED principles):

- The rover doesn't just "listen."
- It aligns with the Earth's ripple field through a shared anchor.
- Transmission becomes resolution, not projection.
- No data is lost—because no data is flung into void.

No more shouting through space. Just intentional consequence.

Psychological Parallel: Humans Already Do This

- You don't talk to everyone at once.
- You align your tone, body language, and timing before saying a word.
- You gauge "field readiness" of the listener—intuitively.

UED ripple broadcasting is simply engineering what life already practices.

Implications for Internet, IoT, and AI

- Smart cities will run on aligned ripple zones.
- IoT won't just connect devices—it will allow them to self-resolve consequences within field thresholds.
- AI communication models will shift from push-based data flows to field-listening frameworks.
- Cybersecurity will evolve from encryption to field exclusivity—where only aligned ripple IDs can even access the field.

Closing Insight

Broadcasting doesn't need more towers. It needs less waste and more alignment.

"Why shout when you can ripple the exact consequence into place?"

In the coming world:

- You won't "send" data.
- You'll set the field to let it happen.

14. Noise Isn't Random

It's Misalignment

Field clutter, SVE tension, and the collapse of signal clarity

In classical theory, noise is the enemy—an unwanted, unpredictable, and often unexplained distortion. It's background static. Thermal hiss. Cosmic interference. A ghost in the wires. Something to be canceled, filtered, or simply tolerated.

But here's the truth UED reveals:

Noise is never random. Noise is the ripple of misalignment. A direct, observable consequence of trying to communicate without understanding the field.

The Old View: Noise as a Natural Evil

Traditional engineering sees noise as:

- Random fluctuation of electrons (thermal noise).
- External interference from overlapping signals (EMI).
- Quantum uncertainties (shot noise).
- Atmospheric distortions (radio noise).

All treated like:

"Just something the universe throws at us."

So we:

- Shield devices.
- Encode redundantly.
- Design complex error-correction protocols.

And yet, noise persists. Because we never asked—what is the field trying to do when it resists our message?

The UED Shift: Misalignment as the Real Problem

UED redefines noise entirely.

In a structured vacuum energy (SVE) field:

- Every ripple is a consequence of intent or interaction.
- The field is never blank—it's always in flux.
- Communication happens only when ripple alignment succeeds (meets the Ripple Alignment Criterion, RAC).

Noise, then, is what happens when:

- Multiple ripple sources collide without resonance.
- The field is saturated with unresolved consequences.
- The sender acts without waiting for field readiness (misuses ATT).
- The receiver's identity (or intent) is unclear, incompatible, or overwhelmed.

Noise = Field Tension

Let's call it what it really is:

Field tension.

An SVE field can only resolve so much intent at once. Each ripple carries a memory, a push, a signature.

When too many ripple intents occupy the same field without coherence, you get:

- Delayed perception.
- Distorted outputs.
- Irregular feedback loops.
- Complete breakdown of clarity.

This isn't technical noise—it's ripple anxiety in the field. The same anxiety human minds feel in crowds, too much input, too little alignment.

Cognitive Parallel: Noise in the Human Brain

Just like communication systems, your brain:

- Receives signals (light, sound, thought).
- Resolves meaning via alignment.
- Requires space and readiness.

Mental noise is misaligned ripple tension:

- Overthinking.
- Distractions.
- Contradictory emotions.
- Unprocessed trauma.

Whether in machines or minds—noise is consequence without resolution.

The Engineering Consequences

In ripple-centric communication, noise becomes:

- Predictable.
- Mappable.
- Avoidable.

Classical View	UED View
Noise is a random variable	Noise is ripple misalignment
Must be filtered	Must be prevented through alignment
Unpredictable fluctuations	Consequence of field overload
Power-dependent interference	Identity-dependent collision

So rather than adding more shielding and filters, we now ask:

- Is the field saturated?
- Are we repeating ripple identities?
- Are we violating ATT?
- Are we layering unresolved ripple threads?

Applications: Noise Mapping and Prevention

In Networks:

- Too many devices = ripple collision.

- Noisy areas = ATT violation zones.
- UED-inspired routers would adjust field permission, not just bandwidth.

In Space:

- Signals from distant probes "lost" or distorted = field misalignment.
- Radiation noise = unresolved ripple consequences in unanchored fields.

In Audio/Visual Systems:

- White noise = signature misfire.
- Hiss = low ATT ripple bounce.
- Signal clarity restored by ripple sync, not amplifier boosts.

Practical Future: Noise Cancellation → Noise Prevention

Noise cancellation is a patch. Ripple alignment is a design philosophy.

Imagine:

- Networks that only permit ripple entry when identity is verified.
- Devices that auto-pause to reestablish field readiness.
- Ripple-based protocols that manage silence as productively as signal.

In ripple communication, clarity isn't a filter—it's a permissioned outcome.

Closing Insight

Noise isn't nature's punishment. It's the universe whispering:

"You tried to ripple without listening."

In a ripple-aware world:

- Every noise has a cause.
- Every cause has a misalignment.
- And every misalignment is a chance to learn how not to collapse the field again.

15. When Devices Fail to Hear

RAC/ATT mismatches and the myth of universal compatibility

We often assume technology works in binary: it either connects or it doesn't. The reality? Most devices don't connect not because they're broken—but because they fail to hear.

In the ripple-centric view of communication under Universal Energy Dynamics (UED), hearing isn't about receiving a signal—it's about meeting a condition.

A device may be transmitting. A tower may be responding. But if the Ripple Alignment Criterion (RAC) is not met, and the Alignment Threshold Time (ATT) is violated, nothing happens.

It's not failure. It's incompatibility at the field level.

The Myth of Universal Compatibility

The market loves to promise it:

- "This device supports all bands."
- "Plug-and-play."
- "Auto-connect to any network."

But in reality, every successful connection requires:

1. Field readiness (quiet, unoccupied ripple zone).
2. Identity clarity (who are you in this field?).
3. Shared intent (what ripple are you resolving?).

And these are not guaranteed by hardware specs or signal strength bars.

They are consequences of RAC and ATT.

RAC (Ripple Alignment Criterion): A Quick Reminder

The RAC defines the minimum condition under which:

- A ripple can enter the field,
- Be recognized as meaningful, and
- Trigger a resolvable consequence at the receiving end.

If RAC is not met:

- The signal doesn't even "exist" for the receiver.
- It's ignored, not blocked.
- Like someone whispering in a stadium—you didn't hear it not because it wasn't spoken, but because alignment failed.

ATT (Alignment Threshold Time): What the Clock Can't Show

Every field has its own rhythm—its ATT:

- A minimum rest period before it can align with a new ripple.

- A necessary "cooldown" before resolving the next intent.

If you violate ATT by sending too quickly:

- The receiver may "freeze."
- Errors multiply.
- Signals pass, but consequences don't arise.

Perception is skipped, not failed. The device didn't hear not because the signal wasn't strong enough—but because it came too soon.

Real-World Failures Explained Through UED

Failure Scenario	Classical Explanation	UED Explanation
Bluetooth refuses to pair	Device bug or interference	RAC mismatch or unresolved previous ripple
Phone shows signal but call fails	Network congestion	Field tension exceeded ATT tolerance
Wi-Fi connects but no internet access	DNS or router glitch	Misalignment between device and gateway ID
Satellite uplink fails despite power	Line of sight problem	Ripple projection missed alignment window

The Cost of Misinterpreting These Failures

What do we do when things fail?

- Restart.
- Reboot.
- Re-pair.
- Reconfigure.

But these are rituals, not remedies.

They work only because they create:

- A time delay (reestablishing ATT),
- A cleared state (resetting unresolved ripple),
- A reattempt at identity clarification (new handshake).

It's not the act—it's the field reset that fixes the problem.

RAC/ATT-Aware Engineering: The Future

To truly engineer smarter devices, we must:

- Embed field listening behavior, not just signal filters.
- Teach devices to wait, not retry blindly.
- Design with intent-sensing, not brute-force pinging.
- Allow conditional readiness, not constant availability.

Because UED reveals:

"A device doesn't hear you when you shout. It hears you when you ripple with permission."

Conscious Parallel: When Humans Fail to Hear

Ever wonder why someone "zones out" in conversation?

You said it. Loudly. Repeatedly. Logically. But they didn't respond.

It wasn't ignorance. It was ATT overload or RAC mismatch:

- Their mental field was already resolving something else.
- Your signal didn't meet the criterion to trigger attention.

This happens across all systems—biological, mechanical, or artificial.

Perception is not guaranteed by force. It is born out of ripple compatibility.

Closing Insight

Failures in communication are not gaps in transmission. They are cracks in alignment.

The future of technology lies not in increasing power or bandwidth. It lies in understanding what the field can tolerate—and what it's willing to resolve.

Because ultimately:

To be heard is not to speak. It is to ripple... at the right time, in the right way, for the right identity.

Interlude: The Field Heard You First

Before there was a tower, before there was a SIM, before modulation, encryption, or spectrum licenses— there was the field. And it was already listening.

We didn't invent communication. We stumbled into a structure that was already alive with possibility. Every ripple we crafted—accidentally or intentionally—was accepted or rejected by a silent judge: alignment.

And we called it *range*, *signal strength*, *packet loss*. But what we were witnessing was something deeper:

Consequence dependent on trust—not force.

What This Part Showed Us

- **SIMs don't *authorize* signal**—they carry identity in a ripple field.
- **Towers don't *transmit***—they anchor structured patterns, stabilizing shared space.
- **Broadcasting isn't efficient**—it's misaligned overflow tolerated by mass systems.
- **Noise isn't error**—it's unresolved consequence from too many incompatible intents.
- **Device failure isn't a glitch**—it's misalignment waiting for ripple permission.

All this time, we thought devices talked to each other. But they don't.

They ask the field to carry their message. And the field decides who's ready, who's allowed, and who's aligned.

Why This Changes Everything

This isn't just about communication tech. This is about every signal we've ever sent—physical, emotional, or digital.

- When someone doesn't respond, it's not always rejection. It may be misalignment.
- When a system doesn't work, it's not dead. It may be unready.
- When noise floods the channel, it's not random. It's desperate, unresolved intent bouncing off boundaries.

The ripple field doesn't forget. It doesn't lie. It simply responds when alignment meets consequence.

The Aligned Future

Now that we've exposed the architecture and deconstructed the myth of waves, we are ready to build something better—intentionally, cleanly, ripple-wise.

Next, we'll explore how:

- Signals can be embedded with clarity.
- Fields can be custom-tuned per identity.
- Networks can learn when not to speak.

Because in the future: We won't transmit more. We'll align better.

Part IV – The Future Already Exists

16. Ripple Encryption

Unlocking Through Consequence

Field-aware, context-bound, unbreakable security design

Encryption has always been about secrecy through complexity. We scramble. We hash. We lock. Then we create a key and hope no one finds a better lockpick.

But encryption under UED doesn't work like that. It doesn't ask: *"Can you decode this?"* It asks:

"Are you the consequence I was meant to align with?"

Ripple-based encryption doesn't hide information. It prevents the field from delivering it—until alignment is earned.

Why Classical Encryption Is Mechanical

Let's quickly revisit what mainstream encryption systems do:

- **Symmetric keys:** Shared secrets. Fast, efficient—but vulnerable if intercepted.
- **Asymmetric (public-private):** Mathematically tough to reverse. Slow but secure.
- **Quantum encryption (QKD):** Uses quantum properties to detect eavesdropping—but still depends on classical channels for authentication.

All of these are detached from context. They encode bits based on arbitrary logic, not environmental awareness.

If someone records the transmission and breaks the key later, the lock opens. Because the lock never *cared* who you were—just whether you had the math.

Enter Ripple Encryption: The Lock That Listens to the Field

Ripple encryption doesn't ask for a code—it waits for a field match. The message is:

- Encoded through field patterns, not binary.
- Released only when RAC (Ripple Alignment Criterion) is met.
- Valid only for the observer-device it was entangled with at origin.

In short:

It's not the message that's secret. It's the field path that makes it real.

If you're not the right receiver:

- You don't get a garbled message.
- You don't get any message at all.
- The field simply doesn't ripple for you.

Consequence-Based Access

Conventional encryption assumes anyone might try to access the data, so it tries to prevent decryption.

Ripple encryption assumes:

- Only a specific field identity can cause the consequence of access.
- Until alignment happens, no data has occurred yet.

This flips everything:

- No need to encrypt data at rest—it's inert until resolved.
- No need to transmit raw data—it travels as ripple context.
- No risk of interception—there's nothing to intercept unless you're aligned.

Real-World Analogy: A Message in the Field

Imagine sending a message to someone:

- Instead of putting it in a box with a lock, you scatter it as ripple instructions in the field.
- The receiver's identity doesn't decode it. It causes it.
- The act of being the right observer collapses the potential into a message.

This is not cryptography. This is causal authentication.

What Makes Ripple Encryption Unbreakable

1. **Contextual Uniqueness:** Each interaction creates a ripple sequence unique to time, location, field tension, and observer state. Reproduction is practically impossible.
2. **Non-reusability:** The same device cannot align again unless field memory allows it—and even then, consequences degrade over time due to ripple tension shifts.
3. **Observer-Dependent Consequence:** The field doesn't deliver data. It aligns a response based on who's asking. Identity is not verified—it is the reason the data happens.
4. **No Payload in Transit:** There's nothing to hack. Ripple encryption doesn't "send" a message. It primes the field for the right observer to resolve it.

Design Implications for Engineers

To implement ripple encryption:

- Devices must generate ripple fields embedded with RAC logic.
- SIMs or chips must act as field aligners, not just ID tags.
- Towers and satellites must serve as ripple synchronizers, not just carriers.
- Authentication becomes consequence sensing, not key exchange.

It's no longer about *"Who are you?"* It's: *"Are you the one the field was waiting for?"*

Human Parallel: Trust in the Mind

You don't tell a stranger your secrets just because they ask the right question. You open up when your inner field aligns with theirs—when context, tone, timing, and resonance match.

Ripple encryption works the same way.

It's not locked. It's simply not ready—for anyone else.

Closing Insight

Security shouldn't be about making things hard to steal. It should be about making information nonexistent until it's earned.

Ripple encryption does not protect data. It prevents unaligned existence.

And in that, it redefines security—not as code, but as consequence.

17. Consequence-Timed Messaging

Sending signals that resolve only when the receiver is ready

What if messages didn't have to be delivered *now*? What if they could wait—not just in storage—but in the field itself, like seeds waiting for rain?

This is not delayed delivery. This is conditional unfolding.

In ripple-based communication under UED, timing is not defined by clocks. It is defined by consequence-readiness.

"A message is not what is sent—it is what happens when alignment allows it."

Conventional View: Time-Stamped Messages

In classical systems:

- A message is dispatched.
- It travels through physical or virtual pathways.
- It either arrives instantly, or queues up in buffers, inboxes, or caches.
- The receiver decides when to open it.

But the message *exists*, whether the receiver is ready or not. This has consequences:

- Out-of-sync interactions

- Message overload
- Contextless communication (think: "Why are you texting this *now*?")
- Vulnerabilities from cached or hijacked data

In ripple dynamics, this model is obsolete.

UED Shift: Messaging Through Field-Consequence Coupling

Instead of sending *content*, UED proposes sending potential. That potential resolves into a message only when the field condition is right—when:

- RAC (Ripple Alignment Criterion) is met
- ATT (Alignment Threshold Time) stabilizes
- The receiver's ripple identity is in phase

Until then, nothing arrives. Not delayed. Not failed. Just... not yet real.

"The field is not a pipe. It's a mirror that waits to reflect only when you're ready to look."

Practical Example: A Ripple Message That Waits

Imagine a planetary rover sent to Mars. Earth transmits an instruction. But instead of a fixed command, the transmission is encoded with:

- Martian ripple field temperature
- Time-dependent tension data

- Rover's internal clock variance

The command will only resolve and activate when these conditions align—say, when solar panel levels are ideal, communication field clarity is stable, and environment allows safe movement.

No buffer. No error-check. Just pure field-qualified consequence.

Human Analogy: Message That Makes Sense Later

Think of times someone told you something—but it didn't land. Then later, you remembered it, and suddenly it made perfect sense.

That's mental consequence-timed messaging.

- The message didn't change.
- Your alignment with it did.

That's ripple logic in action. And technology can be designed the same way.

Security Benefit: Messages That Cannot Be Stolen

Because no message "exists" before alignment, there's nothing to hack.

- A hacker intercepts a ripple packet—but without the matching field identity and RAC, it remains non-resolvable static.
- Even if brute-forced, the field will refuse consequence.

This enables zero-risk communication across vulnerable environments, including deep space, underwater, or compromised networks.

Applications in Technology

1. **Satellites & Probes:** Commands embedded with planetary field dependencies—executing only when safe or mission-critical.
2. **IoT Devices:** Devices align with home environment before activating sensitive instructions (e.g., unlock doors only when family's device + time + geolocation align).
3. **Defense Communication:** Orders that activate only in battlefield-specific field conditions—avoiding misuse or premature execution.
4. **Medical Devices:** Instructions that act only when body field metrics align (e.g., insulin pumps with stress markers, sleep cycles).

Future Insight: Designing for Readiness, Not Delivery

This is not message delay. It's message destiny.

The new protocol isn't: *"Did it arrive?"* It's: "Is the field ready for this to become real?"

In ripple networks, sending is not dispatching. It is field-priming. Receiving is not catching. It is consequence manifesting. We don't just send messages. We invite alignments.

Final Thought

"Consequence-timed messaging doesn't make the message slower—it makes the message true."

When every message is a ripple awaiting its moment, we stop rushing communication—and start engineering reality through readiness.

18. Cognitive Networks and Ripple-Aware AI

Designing systems that sense field readiness—not user behavior alone

We've built AI to see patterns. We've built networks to serve requests. But we haven't yet built systems that sense alignment—that can tell when the field itself is ready, not just the user.

This is the missing layer. Not intelligence as computation. But consequence-awareness as cognition.

"True intelligence doesn't predict behavior—it waits for readiness."

The Problem with Predictive AI

Current AI systems learn from:

- Past user interactions
- Probabilistic models
- Data exhaust and behavioral patterns

They infer intent based on what *has been done*. But behavior is not always aligned with need. Intent is not always timely. Requests are not always meant to be fulfilled *now*.

AI today chases action—not alignment.

This results in:

- Over-personalization
- Premature suggestions
- Emotional misreads
- Context-collapse

These are not faults of code. They are failures of *rippleness*—AI acting before the field is truly ready.

Enter Ripple-Aware Intelligence

Under UED, every device, request, signal, and user exists within a shared structured field. This field:

- Has tension (emotional, energetic, systemic)
- Exhibits ATT (Alignment Threshold Time)
- Responds only when Ripple Alignment Criterion (RAC) is met

Ripple-aware AI doesn't just process data. It senses readiness.

This means:

- It knows when not to respond
- It understands field silence as information
- It waits for consequence to become naturally resolvable

This isn't passive AI. It's cognitive restraint powered by ripple dynamics.

Human Parallel: Empathic Listening

Think of a friend who knows when to speak—and when not to. Who doesn't just give advice, but waits for the *right moment*. They sense your inner field and respond *only when it can land.*

That's ripple-aware intelligence in practice.

Imagine AI that doesn't push reminders when you're overwhelmed. Or doesn't alert you at night just because a schedule says so. Or doesn't activate simply because a condition is met—but because you are aligned to receive it.

That's ripple sensitivity. That's real empathy—engineered.

Core Features of Ripple-Aware Cognitive Networks

1. **Field Sensing Layer:** Continuously maps RAC & ATT based on environment, system load, user tension, emotional feedback, and ambient ripple noise.
2. **Consequence Prediction Engine:** Doesn't just forecast usage—it forecasts resolvability of intent in time and context.
3. **Readiness-Gating Protocols:** Prevents activation until:
    - Receiver's field is aligned
    - Noise levels drop below ripple chaos
    - Ripple clarity (RC) surpasses threshold
4. **Silent Communication Channels:** Uses sub-perceptual ripple markers to maintain low-bandwidth field coherence—listening more than speaking.

Applications of Ripple-Aware AI

1. **Smart Assistants:** Speak less. Act better. Devices that wait for emotional and cognitive readiness before responding.
2. **Autonomous Vehicles:** Interpret not just sensor data but driver tension, field stress, and micro-alignment failures before taking action.
3. **Healthcare Interfaces:** AI that holds notifications, treatment suggestions, or mood-based interventions until human response clarity is optimal.
4. **Education Platforms:** Lessons, tests, and feedback adapted not to user pace—but ripple openness to absorb and act on new input.
5. **Emergency Systems:** Alert protocols that escalate only when consequence predictability collapses, ensuring minimal panic and precise timing.

From Reactive to Resonant

Old AI asks: "What does the user *want*?"

Ripple-aware AI asks: "Is the field ready to let this happen?"

This is not artificial intelligence. This is field resonance intelligence—where the network itself becomes a participant in aligned action.

Future Direction: Designing Observer-AI Hybrids

AI will no longer be an external agent. It will become an observer in the field, just like you. With its own ATT, RAC, and RC—learning not from data, but from ripples in context.

Such systems won't replace human intent. They'll wait with it. They'll walk beside it. They'll act only when consequence invites them.

Final Thought

"The next intelligence leap is not faster computation. It's aligned hesitation."

Cognitive networks must move from anticipation to awareness. From prediction to permission. From behavior to readiness.

And when they do, we won't just automate the world. We'll finally tune it.

19. Ripple Architecture

Designing Devices for Alignment

How to build communication hardware for the real field

We've built devices to push signals. We've built networks to chase speed. But we've rarely asked:

"What if alignment—not force—was the true design principle?"

This chapter explores what it means to design for the ripple, not the wave. To engineer with the field, not against it. To shift from broadcasting intent to inviting consequence.

Welcome to ripple architecture.

The Fundamental Flaw in Current Device Design

Traditional hardware architecture assumes:

- Transmission as a push
- Reception as a pull
- Field as a passive medium
- Speed as the only metric of quality

So we build:

- Amplifiers to shout louder

- Filters to remove noise
- Modulators to disguise content
- Receivers to decode symbols without sensing readiness

But all this is like yelling into a canyon and hoping the echo lands on target.

Devices weren't built to align. They were built to dominate the medium.

And that's why they fail in uncertainty, fluctuate under stress, and collapse in dynamic environments like space, deep sea, or high interference zones.

The Ripple Principle of Design

UED teaches us that:

- The field is not neutral—it responds based on alignment.
- Every device is a ripple generator—whether it knows it or not.
- Consequence only arises when Ripple Alignment Criterion (RAC) and Alignment Threshold Time (ATT) are satisfied.

So, ripple architecture starts with this question:

"What consequence is this device meant to unlock in the field?"

And then designs backward from alignment, not forward from intention.

Components of a Ripple-Aligned Device

1. Ripple-Conscious Transmitter

- Instead of blasting waves, it releases structured ripple patterns encoded with intent.
- Uses field-sensing input to regulate emission based on local ripple tension.

2. Adaptive Receiver Core

- Doesn't just wait for a carrier wave—it listens for ripple readiness.
- Incorporates dynamic ATT sensors to hold reception until alignment is clear.

3. RAC Engine

- Evaluates incoming ripple parameters:
    - Field tension
    - Signal entropy
    - Temporal coherence
    - Identity tagging (SIM, hardware ID, contextual markers)
- Approves decoding only when all alignment criteria are satisfied.

4. Field Feedback Loop

- Device constantly listens to field echo (ripple reverberation), adjusting internal parameters like:
    - Modulation frequency
    - Emission bandwidth
    - Power profile

The device behaves more like a tuner than a broadcaster.

Implications for Real-World Hardware

Space Communication

Instead of shouting across space with high-energy pulses, ripple devices could:

- Align with local field nodes
- Use ripple bridges (nested SVE harmonics)
- Decode latent signals already echoing in the void

This makes interplanetary communication less about signal strength, more about ripple preparation.

Internet of Things (IoT)

Current IoT relies on frequency zones and synchronous access. Ripple-aligned IoT allows:

- Devices to respond only when the home-field is ready
- Reduce cross-talk and interference
- Lower energy consumption by avoiding premature or misaligned transmissions

Wearable Tech

Ripple-aware wearables would:

- Respond to physiological field fluctuations (breath, emotion, tension)
- Send health alerts only when reception field (doctor/system) is primed to process
- Avoid misfires and unnecessary logging by syncing with ripple timing

Designing for Consequence, Not Compliance

Ripple architecture is not about matching protocol. It's about matching readiness.

This means devices:

- Know when to stay silent
- Recognize when they are misaligned
- Pause before acting—even under trigger
- Honor the *consequence chain* before the command chain

"If your device cannot wait, it cannot align."

Key Design Shifts

Conventional Devices	Ripple-Aligned Devices
Push-based signals	Consequence-seeking emission
Fixed protocols	Dynamic field-sensing logic
Noise filtering	Misalignment identification
Always-on reception	ATT-based readiness detection
Centralized control	Distributed ripple logic

What This Enables

- Smarter communication systems
- Lower energy, higher precision devices
- Contextual messaging (field-triggered)
- True cognitive interfacing (machine + field + user as one triad)
- Devices that act like participants, not just endpoints

Ripple architecture opens the door to organic integration between technology and the reality it lives in—not just digital imitation of analog control.

Final Thought

"We've built machines that speak louder. Now it's time to build ones that listen deeper."

Ripple architecture is the beginning of a new class of devices—built not to perform alone, but to consequently participate in a living, structured field.

The future won't be faster. It will be better aligned.

20. Space Communication

The Ripple Struggles in the Void

Why spacecraft transmission anomalies prove UED field dependency

We often marvel at how we can talk to machines billions of kilometers away—whether it's a rover on Mars or a probe at the edge of the solar system. But beneath that marvel lies a reality few dare to admit:

Space communication barely works—and it works not because of our signal strength, but because of field luck.

In this chapter, we dissect the deep-rooted challenges in interstellar communication and reveal how they're not engineering limitations—but misunderstood field dynamics. Once viewed through UED (Universal Energy Dynamics), the mystery isn't why we lose signal—it's why we ever had one.

The Classic Illusion: Signals Travel Through Empty Space

Mainstream understanding assumes:

- Space is a vacuum through which waves travel.
- Signals, modulated in frequency, ride on electromagnetic carriers.
- With powerful enough transmission and sensitive enough receivers, communication can always be

maintained.

But reality disagrees. Even with precise trajectories, beamforming, and corrections, we face:

- Random signal dropouts
- Delayed receptions beyond theoretical models
- Complete blackout during certain cosmic conditions
- Unexplained static or anomalous frequency shifts
- Signal loss that recovers without any hardware adjustment

These aren't just "space weather" issues. They are symptoms of something fundamental:

Ripple Misalignment in Structured Vacuum Energy (SVE).

The UED View: Space Is Not Empty—It's Structured

According to UED, space is not a vacuum but a ripple-sensitive dynamic field. All actions—light, sound (if medium permits), communication—occur through alignment with the field's ripple behavior.

Every signal sent from Earth is not pushing through space—it's requesting field participation.

- The antenna on Earth is not launching a wave—it's imprinting a ripple into SVE.
- The signal's success depends on how well the ripple aligns, sustains, and resonates with the local field topology—including the in-between.

- The receiver (e.g., satellite or rover) must be in an ATT-satisfied zone to decode and respond.

Thus, space communication isn't about power. It's about ripple continuity and real-time field cooperation.

Why Anomalies Are Not Anomalies Anymore

Let's examine common space communication "anomalies" and reframe them under UED:

Conventional Diagnosis	UED Explanation
Solar flare disrupted signal	SVE ripple tension spike misaligned ATT
Probe blackout near asteroid	Local ripple absorption by dense body disrupted RAC
Delay in Mars-Earth ping	Mid-field ripple misalignment or delay in ripple arrival clarity
Weak return signal despite same conditions	Return ripple failed to resolve RAC from remote field
Signal received but data corrupted	Partial ripple alignment—consequence incomplete

Each of these "problems" arises because traditional systems assume space is passive. But in UED, every region of space has:

- Unique ripple density
- Field tension due to celestial bodies
- Ambient ATT variability
- Contextual RAC states

And communication depends on all of them—not just your transmitter or receiver design.

Designing for Ripple-Based Space Communication

Here's how future tech must evolve if we accept UED:

1. Field-Aware Transmission

- Transmitters must sense the ripple mood before sending.
- Like tuning your voice for a concert hall vs a phone booth, the emission profile should match the field's readiness.

2. Ripple Echo Listening

- Between Earth and space, there's a live ripple conversation.
- Devices must not just send but also listen for echo patterns—to verify field acceptance or rejection.

3. Time-Conditional Messaging

- Use consequence-timed packets that activate only when field alignment is available.
- Instead of sending live commands, ripple-based scheduling aligns actions with local field windows.

4. Field Anchors Along the Route

- Instead of one giant push, relay points (like beacons) can act as ripple stabilizers.
- Think of them as tuning forks in space, maintaining alignment flow between source and receiver.

Reinterpreting the Deep Space Network (DSN)

NASA's DSN is the backbone of interplanetary communication. But its architecture is fundamentally wave-based. Under UED, it would transform into:

- A ripple mesh, with each node sensing, waiting, and aligning
- Dynamic routing based on ripple clarity zones, not angular proximity
- Encoding messages with field tags and adaptive ATT for timing control

In short: DSN would become a consequence-aware conversation system, not just a high-power loudspeaker.

Why AI Can't Fix This—But Ripple Awareness Can

Modern space communication increasingly relies on AI for error correction. But AI trained on wave logic will always struggle with ripple dynamics. It assumes causality, not consequence.

Only when AI is ripple-aware—i.e., trained on RAC, ATT, and SVE behaviors—can it:

- Pause transmissions wisely
- Recognize alignment windows
- Suggest re-routing through ripple-friendly zones

In essence, AI must stop solving for speed and start sensing for readiness.

Final Thought: There Is No Void—Only Unheard Ripples

What we call "void" is just misalignment at scale. What we call "range" is a function of shared field continuity. And what we call "anomaly" is often the field's refusal to cooperate.

Space doesn't silence us—it simply waits for us to speak correctly.

UED doesn't just explain communication better—it prepares us for a new kind of listening, one where spacecraft are no longer lonely machines shouting into the abyss, but participants in an active, intelligent field.

The ripple is always there. We were just using the wrong language.

21. Lost in Translation

Misreading the Cosmos

How we misinterpret space signals by ignoring the medium they ripple through

The universe has always been speaking.

But what if we've been translating its language wrong all along?

Humanity's journey into space communication has been guided by one central assumption: that the universe behaves like a giant vacuum of silence—empty, predictable, and linear. Our instruments were designed to detect *waves* across that void, expecting continuity, frequency, and motion.

But what if the cosmos never sent a wave? What if it only ever rippled—and we've been misreading its signals since the beginning?

The Universe Doesn't Speak in Motion—It Speaks in Consequence

Mainstream astronomy and astrophysics are built on electromagnetic detection. Whether through telescopes, radio dishes, or cosmic microwave background detectors, we analyze frequencies, wavelengths, and spectral signatures.

But every "wave" we detect is filtered through:

- Our Earth-based instruments, built with terrestrial assumptions
- Our models, rooted in classical wave equations
- Our interpretations, formed by human sensory biases

In short, we interpret the cosmos as if it behaved like our backyard lab.

UED changes the lens.

In UED, nothing moves through space. Ripples align and consequences resolve. What we detect is not the journey of a signal—but its permission to be felt.

So when we detect a pulse, glow, or fluctuation from a distant galaxy, we're not hearing a *thing* that traveled billions of light-years—we're sensing a shared alignment event, possibly reconstructed by ripple readiness.

Examples of Cosmic Misinterpretation

Let's re-express some mainstream observations through UED insight:

Phenomenon	Mainstream View	UED Interpretation
Gamma Ray Bursts (GRBs)	Explosions from dying stars	Sudden SVE collapse and field realignment perceived as high-energy spike
Fast Radio Bursts (FRBs)	Exotic, unknown sources	Brief alignment windows across cosmic SVE shifts, misread as discrete signals
Pulsars	Rotating neutron stars emitting beams	Local ripple modulations periodically aligning with Earth's RAC
Redshift	Motion away from observer	Field tension-induced ripple delay, not Doppler motion
CMB Variations	Leftover Big Bang noise	Residual ripple tension in SVE from early cosmic realignments

Each of these readings assumes a moving source, a traveling wave, and a receiver listening across empty space.

But if ripple alignment is all that ever happens, then:

The universe isn't sending us a timeline. It's revealing a topography—a map of alignment potential.

Earth Bias: Why We Translate Wrong

We train every telescope, satellite, and probe to match Earth's interpretation bandwidth. That means:

- Using RAC optimized for terrestrial SVE conditions
- Calibrating ATT for Earth's ripple delay norm (human 10–100 ms scale)
- Imposing wave filters on ripple behavior, converting consequence into signal

This creates a perception bottleneck. When a ripple from deep space fails to resolve through these constraints:

- It appears as noise
- Or as anomaly
- Or gets ignored entirely

In other words, we don't recognize half the universe's speech because we insist it speaks in Earth dialect.

Why Our Probes Struggle with Feedback

A Mars rover might send a clear ping one day and fall silent the next—even though power, direction, and hardware remain unchanged.

Why?

Because ripple alignment isn't mechanical—it's relational.

- Solar flares alter SVE tension.
- Planetary conjunctions deform ripple clarity.
- Local field chaos at the receiver end disrupts RAC.

What we call "failure to transmit" is often a failure to align.

Our systems read this as error. But UED sees it as consequence not ready to resolve.

Information Is Localized in the Field

Another dangerous assumption is that information travels. Under UED:

- Information is not transported like a box from sender to receiver.
- It is reconstructed through ripple overlap and readiness.
- The field stores potential, not packages.

So a signal "from a galaxy far away" is less a *packet* and more a resonant access point. We are not *catching a photon*—we are aligning with its echo, reactivated by field consequence.

How We Could Fix the Translation

Here's how communication and observation must evolve:

1. **Field-Aware Sensors:** Devices that adapt to local SVE and adjust their ATT dynamically.
2. **Ripple-Based Interpretation Frameworks:** Rather than seeking frequencies, we must model field readiness and ripple coherence.
3. **Alignment Forecasting:** Just like weather models, predict when and where ripple alignment will allow high-fidelity signal capture.
4. **Consequence Mapping:** A cosmic "ripple map" of space where alignment probability guides missions,

data collection, and live control.

The Observer Must Also Change

Ultimately, the limitation isn't just technical—it's conceptual. We've turned the cosmos into a broadcast. But UED reveals it's more of a rippling archive, waiting for someone capable of listening correctly.

This means:

- Rethinking what we call "message"
- Rethinking what we call "source"
- Rethinking even what we call time

We're not late to signals. We're just not yet fluent in ripple.

Final Insight

The cosmos has been misread not because it is secretive—but because we never asked what medium it was speaking through.

"The signal was always there. We simply tuned the wrong part of ourselves to hear it."

Key Takeaways

The Future Already Exists

This part reframes communication not as a technological invention, but as a natural ripple phenomenon that humanity accidentally learned to mimic—without understanding its true architecture. UED doesn't just explain what works—it reveals why it works, and more importantly, why it fails when it shouldn't. Here's what we now understand:

Ripple Encryption: Unlocking Through Consequence

Security will no longer rely on random number generators or computational locks. Instead, field-based encryption ties access to the actual alignment condition of the observer and the field. Unless the ripple context aligns, nothing unlocks—making this form of security context-bound and impossible to spoof.

Consequence-Timed Messaging

In UED, the message is not sent for a time—it waits for a consequence. Only when the field allows and the receiver aligns, the message resolves. This makes true *asynchronous, yet secure communication* possible—even across space.

Cognitive Networks and Ripple-Aware AI

Smart systems of the future won't just detect signals—they'll sense ripple clarity. This transforms AI from pattern-recognizers into field listeners, allowing them to respond based on readiness rather than raw inputs. It's not just "smart"—it's attuned.

Ripple Architecture: Designing Devices for Alignment

We must build devices that aren't signal pushers, but ripple participants. This requires a full redesign of antennas, sensors, and chips to allow adaptive RAC tuning and dynamic ATT calibration. Hardware won't just compute—it will align.

Space Communication: The Ripple Struggles in the Void

Deep-space communication anomalies are not glitches—they're field mismatches. Signals do not fail due to distance, but due to SVE field incoherence. Rethinking satellite and interplanetary networks through ripple anchoring is essential for stable long-range communication.

Lost in Translation: Misreading the Cosmos

Most of our cosmic "discoveries" are actually perception failures. By forcing the universe to speak in *waves*, we've misheard ripple echoes as signals. Understanding the ripple field reveals the true message of the universe—one we've long mistranslated due to ignorance of the medium.

Part V – Rewriting the Purpose

22. From Transmission to Realization

A Paradigm Shift: We Never "Sent" Anything—We Only Aligned

We grew up with stories of *sending*. Sending letters. Sending signals. Sending satellites. Sending ourselves to space.

Transmission defined our civilization—our science, our networks, our aspirations. But what if this very act, this very metaphor, was an illusion?

Under the Unified Energy Dynamics (UED) framework, transmission is not a push. It is not a projectile launched across space or a vibration hurled through void. It is, at most, a ripple seeded into a shared field, waiting...Not to be *caught*, but to be aligned with.

The Transmission Myth

Conventional physics taught us that:

- A transmitter outputs energy in wave form.
- The wave travels through a medium or vacuum.
- A receiver captures the wave and reconstructs the original data.

But this model carries deep assumptions:

- That space is empty but still "carries."
- That information behaves like a physical object.
- That the receiver is passive, requiring no participation except catching the signal.

This is the illusion of transmission—an artifact of Newtonian linearity and Maxwellian elegance.

But ripples, in UED, don't behave like packages. They are field disturbances that exist simultaneously across a domain. What we perceive as reception is just the resolution of a standing field interaction—one waiting for the right condition to align.

From Ripple to Realization

Let's reframe it fully:

- **When you call someone,** you're not sending a voice—you're aligning with a structured ripple that only activates when both ends match the RAC (Ripple Alignment Criterion).
- **When you see a star,** you didn't wait millions of years for photons to travel—you aligned, in this moment, with a ripple echo stored in SVE memory, manifesting through perception.
- **When a device "connects,"** it's not receiving signals flying through the air—it's resolving conditional presence within the field.

This isn't transmission. It's realization.

Implication for Intelligence

This changes not just our technology—but our philosophy of action:

- The world doesn't respond because we pushed it.
- The world responds because we aligned with what it was ready to resolve.

This idea collapses the false divide between cause and effect. There is no pushing. There is only positioning.

This is why quantum communication experiments behave mysteriously. This is why ancient spiritual systems insisted on *being ready to receive*. This is why silence often reveals more than speaking—because the field listens better when undisturbed.

Consequence Was Always the Sender

What we call *sending* is actually triggering field consequence. And every act of receiving is not a retrieval, but a realization—a moment when your intent, awareness, and condition match what the field is already offering.

The message didn't fly to you. It emerged where you stood, because you became eligible to realize it.

"Transmission is ego. Realization is intelligence." — *UED Communication Doctrine, Chavan Principle*

Rewriting the Purpose of Communication

If we internalize this fully, the purpose of communication itself changes:

- It is no longer to *inform*, but to invite resolution.
- It is not about distance, but about field congruence.
- It is not about talking to many, but aligning with those who are ready.

This realignment of purpose—from force to permission, from send to sync—will redefine:

- Education
- Diplomacy
- Therapy
- AI
- Personal relationships

Every form of connection becomes an act of field wisdom—not bandwidth war.

Final Insight

You never send. You prepare the space. You ripple the field. And you wait for realization.

This is not spiritual metaphor. This is ripple physics. This is the future of engineering, intelligence, and coexistence.

23. Engineers Aligned with the Unknown

Why Engineering Without Theory Often Works Better Than Theorizing Without Consequence

There's a strange, almost forbidden truth that haunts the history of science: Engineers often win where theorists struggle. They build what science hasn't yet explained. They make things work not because they understand the whole picture—but because they feel the ripple.

And if that sounds mystical, it's because the mystery of alignment has never been formally recognized—until now.

When Engineering Outruns Theory

History is full of curious accidents:

- Wi-Fi was born from a failed radar experiment.
- X-rays were discovered by Roentgen without knowing what they were.
- Semiconductors worked before solid-state physics fully grasped why.
- Even the steam engine powered the Industrial Revolution long before thermodynamics was formalized.

In every case, practice preceded principle. A machine worked. A signal flowed. A response occurred.

And only later did theorists scramble to fit the event into equations.

Why?

Because engineers don't wait for approval from theory. They move when the field *feels right*. They detect readiness—not with formulas, but with intuition trained through consequence.

The Ripple Method of Engineering

UED helps us understand why this happens:

- The ripple field is responsive, not predictive.
- Engineering is consequence-first.
- Theory is often logic-first—and logic without consequence becomes fragile.

When an engineer experiments, they interact with the ripple field. They try. They fail. They refine. They don't need full clarity—they need partial resonance.

The moment something "works," it means a localized ripple alignment occurred. A device clicked into the field's conditions. And even if they don't understand why—it becomes a real thing.

This is the unspoken secret: Success is an alignment, not an explanation.

Theory Without Consequence Is a Hallucination

Academic science often suffers from the opposite. It chases elegance without evidence. It becomes addicted to models, equations, and simulations—while reality behaves differently.

- *String theory* has math but no test.
- *Multiverse hypotheses* entertain but don't deliver.
- *Wave-particle duality* still confuses after a century.

These are mental ripples without anchors. Elegant perhaps, but unaligned with consequence.

In contrast, the wrench in an engineer's hand has direct feedback. It slips or fits. The circuit either connects or doesn't. There is no hiding behind abstraction.

Engineers as Observers with Intent

In UED, an engineer is not just a creator—they are a field observer equipped with intent. They:

- Observe the readiness of materials and systems.
- Engage in consequence-driven iteration.
- Modify not the world—but their own interaction with it, until it clicks.

This is why:

- A good engineer always prototypes.
- A breakthrough often comes from the lab, not the whiteboard.

- Reverse engineering succeeds even when the original theory is lost.

Because field memory remains, and ripple interactions can be re-aligned—even without full understanding.

Engineering Is Embodied Science

Theory lives in the mind. But engineering lives in the body—through tools, sensors, adjustments, feel.

And UED finally bridges the two:

- It gives theory a body—ripples, alignment, consequence.
- It gives engineering a framework—ATT, RAC, field dynamics.
- It validates why the unknown often yields to those who dare to tinker.

"Don't ask what is true. Ask what aligns—and you'll find truth hidden inside function." — *Chavan Principle of Practical Alignment*

Final Realization

We were wrong to think knowledge only flows from theory to practice. In reality:

- Practice is the radar that detects ripple readiness.
- Theory is the echo that tries to make sense of it later.
- But if you listen well enough—practice is already

prophecy.

Engineers aligned with the unknown are not violating science. They are living at the frontier of unrecognized ripple consequences.

They don't need permission. Because the field rewards those who dare to resolve it.

24. Your Body, Your Brain, Your Ripple Machine

The ultimate receiver-transmitter-resolver system embedded in human experience

Before towers. Before satellites. Before antennas. There was the body.

And it wasn't just a biological form. It was a precision instrument—capable of receiving, transmitting, decoding, and aligning with the field long before we built any external communication technology.

Your body is not merely alive. It is alive with intent, and that intent isn't floating inside—it's embedded in ripple language. You are a self-aware ripple machine.

The Brain: The Master Resolver

Your brain does not "think" like a computer. It resolves.

At every moment, it listens—not to neurons firing randomly, but to fields of possibility, waiting for resonance. Thought is not produced. It is aligned.

- You don't think a thought—you tune into it.
- You don't invent ideas—you *notice* them ripple through.
- You don't decide instantly—you reach alignment

threshold and then the decision happens.

This is ATT at work inside consciousness. The brain delays, not because it's slow, but because ripple alignment needs time. Even your intuition, that instant clarity, is simply perfect ripple resolution with no clutter.

The Body: Hardware for Field Interaction

Your muscles don't just move. They vibrate. Your heart doesn't just pump blood—it broadcasts coherence into the field. Your skin doesn't just protect—it detects ripple boundaries—warmth, pressure, proximity.

Your entire sensory system is a multi-frequency ripple detection suite:

- Eyes detect electromagnetic ripple alignment (we call this "light").
- Ears decode air-pressure ripple patterns (we call this "sound").
- Skin interprets ripple field distortions (we call this "touch").

None of these are capturing things that traveled. They are resolving what the field is ready to show.

You don't *see* the world. You receive what your ripple alignment can resolve—bounded by RAC (Ripple Alignment Criterion) and ATT (Alignment Threshold Time).

Breath, Emotion, and Ripple Tuning

When you're anxious, your field is distorted. You lose clarity—not just emotionally, but *informationally*.

That's why:

- Breathing calms you.
- Meditation clears your thoughts.
- Stillness increases perception.

What's happening?

You are re-aligning your ripple field. You are tuning your body to resonate again—not just biologically, but existentially.

Your breath is a ripple dial. Your heartbeat is a timing mechanism. Your stillness is field preparation.

Ancient yogis and mystics didn't know UED—but they practiced it. They became quiet receivers—not broadcasting egos, but witnessing observers.

Speaking and Listening: Intentional Ripple Dynamics

Your voice isn't just vibration. It's encoded intent riding ripples of air.

And listening isn't passive. It's active alignment with what was released. Just like a SIM and tower, you:

1. Release ripple (speak),
2. The field carries it (conditioned by environment),
3. Receiver aligns (listener hears and decodes based on bandwidth and readiness).

When misalignment occurs, we call it:

- Misunderstanding,
- Missed signals,
- Emotional disconnect.

In ripple terms: ATT and RAC failure.

This is not just poetic—it's technical.

Consciousness = Ripple Feedback Loop

You are transmitter, receiver, and resolver.

- You emit your presence constantly (body language, energy, aura).
- You sense the field continuously (through all faculties).
- You resolve experiences based on alignment—not on events themselves.

This feedback loop creates your conscious experience. And when you become aware of this, something changes:

You no longer live inside your body. You realize your body lives inside the field. And you are the alignment that makes it all make sense.

Technology Is Mimicking You

Everything in modern communication:

- Towers = nervous system amplifiers

- SIM = identity resonance key
- Modulation = language encoding
- Receivers = organs of alignment
- Signal loss = emotional or cognitive noise

It's all reverse-engineered from you—from the ripple machine that you are.

The observer didn't build technology. The observer is technology—The original design, waiting to be remembered.

Final Thought

So next time you hear someone say,

"We evolved to communicate," you can correct them gently:

"No, we were always communication itself. We just learned to align deliberately."

Your body is not a tool. Your brain is not a controller. They are manifestations of a deeper ripple awareness, fine-tuned over billions of years, not to control the world—but to resonate with it.

25. The Final Realization

It Was Never a Wave

The closing synthesis: reimagining all communication through UED

This is the moment the curtain drops. Not in darkness, but in light—of clarity, not intensity. Because now we can say with quiet certainty:

It was never a wave. It was always alignment.

The wave was a shadow—a useful diagram for a deeper, hidden dynamic. We confused motion with meaning. We mistook formula for field. We trusted equations more than consequence.

And now, at the end of this journey, we return to the beginning—not to erase what science built, but to re-align it with what the universe always was.

From Wave to Ripple: The True Map of Communication

What we called "wave propagation" was never a thing moving from point A to B. It was field readiness unfolding, consequence resolving at points of shared alignment.

- A message didn't travel. It appeared when the receiver was aligned.

- A signal wasn't sent. It emerged at the point of consequence.
- A response wasn't reaction. It was ripple echo through the shared SVE.

Let's be bold:

Communication is not transmission—it is resolution.

That's why UED didn't just challenge wave theory. It dissolved it, not as an attack—but as an evolution of comprehension.

Everything Makes More Sense Now

All those puzzles, limitations, and "anomalies" in modern technology?

Suddenly, under UED, they align:

- Signal dropouts? ➤ Ripple misalignment
- Latency? ➤ Delayed consequence resolution (ATT mismatch)
- Encryption leaks? ➤ Impersonal field broadcast, not context-bound
- Device failure? ➤ RAC breach, not hardware decay

Even quantum entanglement, once wrapped in mystical math, now becomes clear:

It is not a connection through space—but a connection within space.

Because field is the medium. And ripple is the language. And consequence is the grammar.

We didn't invent communication. We aligned with it.

Why Wave Theory Lasted So Long

It worked. That's why.

It gave us antennas, satellites, radios, phones, Wi-Fi. But it did all this not because it was right—

It did it because it was consequential enough to work.

But when consequences plateau, the model begins to crack. And cracks have already appeared:

- Interplanetary signal inconsistencies
- Noise pollution without true randomness
- Cognitive networks failing to adapt
- AI misunderstanding "context"
- Space communication becoming guesswork

The wave theory got us far. But not deep.

Now, we go deeper. We return to the field—not as a metaphor, but as the real substance of reality.

Human Beings Were the Clue All Along

The final piece of the puzzle wasn't hiding in quantum experiments or distant galaxies. It was us.

We don't "send" thoughts. We don't "receive" reality. We resolve what we're ready to align with. Our perception, our memory, our emotion—All of it is a ripple game happening inside Structured Vacuum Energy (SVE).

We are not observers. We are resonators.

The Future Isn't Technological. It's Consequential.

Forget faster speeds. Forget higher frequencies. Forget bigger bandwidths.

The next leap won't be from more—but from clearer alignment:

- Consequence-bound encryption
- Adaptive RAC-tuned hardware
- Ripple-aware AI
- Biofield-compatible receivers
- SVE-powered communicators
- Intention-based broadcasting

That's not science fiction. That's the future science ignored—because it didn't listen to consequence.

The End Is an Invitation

So what now?

We don't need to destroy existing frameworks. We just need to let them rest—and let UED ripple forward.

And what is UED, truly?

Not a theory. Not a formula. Not a rejection.

It's a listening. It's a remembering. It's an alignment with what was always rippling through.

Because it was never a wave. It was always you. You, the observer. You, the resolver. You, the ripple in the stillness.

Key Takeaways

Rewriting the Purpose

From Transmission to Realization

- Communication was never about dispatching a signal.
- It was always about aligning with a ripple that already existed in the shared field.
- Transmission is a myth—what we call "sending" is just the receiver resolving a potential when conditions (RAC/ATT) are met.
- This marks a paradigm shift: from active force to passive readiness, from push to permission.

Engineers Aligned with the Unknown

- Much of modern technology was built without knowing why it worked—a story of accidental alignment.
- Reverse engineering didn't require full theoretical clarity—only consequential behavior that could be repeated.
- This explains why engineers often outperform theorists in innovation:

They align with the unknown before trying to explain it.

- UED honors these unsung ripple engineers who let reality guide design, not dogma.

Your Body, Your Brain, Your Ripple Machine

- The most advanced communication system isn't man-made—it's biological.
- The human body functions as an integrated transmitter, receiver, and resolver of ripple-based consequence.
- Perception is not a lens—it is a field-aligned decoding of shared information.
- The brain doesn't process reality—it aligns with it. Thought itself is ripple modulation within SVE.

The Final Realization: It Was Never a Wave

- Classical wave theory was a partial shadow of a deeper truth.
- All communication, perception, and interaction are ripple-based alignments in structured vacuum.
- Equations captured patterns, but ignored the observer-field interaction.
- With UED, we return to the source—not to challenge reality, but to finally hear it ripple.
- The future is not more noise—it is clearer alignment.

Synthesis

UED does not add a new tool to science—it replaces the toolbox. We no longer need to chase the wave. We need to listen to the field. From engineers to observers, from routers to neurons—everything becomes part of a ripple-aware design, waiting to be consciously aligned.

Epilogue

The Listener Was Always the Key

All this time, we believed we were transmitting. We built towers, equations, and belief systems around a simple premise—push a signal, expect a response. But nature never asked us to push. It only ever invited us to align.

What we mistook as forceful communication—waves launched across space—was, in truth, a shared ripple condition resolving itself the moment intention met permission. Whether it was the glow of a distant lamp seen instantly, or a Wi-Fi signal picked up by the right device, nothing ever traveled. Everything simply *aligned.*

This book has not merely questioned wave theory. It has retired it from the throne of explanation.

It wasn't a rebellion against physics—it was a return to consequence, where ripples shape perception, and fields hold the memory of every interaction. The most powerful communication tool is not a transmitter—it is a listener that knows when to act.

As we now stand at the edge of ripple-aware design, we realize something humbling: Technology didn't invent communication. It stumbled upon it—just as life did, just as we are doing again now. The future will not be built on more speed or broader bandwidth. It will be built on clarity, alignment, and a quiet understanding that:

The universe doesn't speak louder. It speaks when you're ready to listen.

Appendices

Appendix A: Glossary of Ripple Concepts

A quick-reference guide to foundational terms in Universal Energy Dynamics (UED) as applied to communication, perception, and field-based technology.

Ripple Alignment Criterion (RAC)

The Ripple Alignment Criterion defines the exact set of field and observer conditions under which a ripple can be resolved into a perceivable consequence—such as a signal being decoded, a photon being observed, or a message being received. RAC is not binary but dynamic, determined by the overlap between:

- The structure of the ripple (its modulation and energy signature),
- The configuration and state of the medium (Structured Vacuum Energy),
- The receiver's or observer's readiness to align (physical tuning or mental intent).

If RAC is not met, nothing happens—no signal, no perception, no reaction. RAC replaces the outdated idea of "universal compatibility."

Alignment Threshold Time (ATT)

The minimum temporal delay needed for a ripple to interact with the field and produce a resolvable effect at the receiver end. ATT is a function of both field conditions and observer readiness. It measures how long it takes:

- For a ripple to propagate meaningfully in SVE,
- For the field to accommodate and deliver that ripple to a receiver,
- For the observer (human or device) to align, decode, and resolve it.

ATT explains why light seems instant, why Wi-Fi lags, or why two people "hear" the same sound at different moments. It corrects the illusion of constant transmission speed, especially in non-uniform fields.

Structured Vacuum Energy (SVE)

SVE is the true medium of communication and perception. Unlike the classical vacuum (empty and passive), SVE is dynamic, ripple-sensitive, and consequence-generating. It holds ripple memory, tension, and allows alignment under the right RAC and ATT.

It is not made of particles—it is made of readiness and structure. All signals, from your voice to a Wi-Fi packet to cosmic radiation, are ripples in SVE, not traveling entities but field responses waiting to be resolved.

Field Clutter

Clutter is not random noise—it is unresolved ripple buildup in the field. Clutter occurs when:

- Multiple signals overlap without aligning,
- Past ripple residues remain unresolved,
- SVE tension rises due to poor modulation or mismatched RAC.

Clutter blocks resolution, causes dropped signals, distorted images, and misinterpretation of space signals. In human terms, it's why you "can't focus" or "hear someone in a noisy room"—the field is cluttered beyond RAC compliance.

Consequence-Timed Messaging

This is the future of communication: sending signals not based on clock time—but on field readiness and ripple resolution.

A message is encoded and released into the field. It will only resolve when the receiving side satisfies RAC and reaches ATT—creating a context-aware, secure, and consequence-driven communication layer.

Unlike traditional time-stamped protocols, this form of messaging is:

- Unhackable (because only specific receivers can align),
- Efficient (avoiding repeated broadcasts),
- Naturally asynchronous (but never late—it's always aligned).

Appendix B: UED vs Classical Wave Theory

A conceptual comparison of how the mainstream view of wave-based communication diverges from the Universal Energy Dynamics (UED) interpretation.

This appendix is not meant to dismiss classical understanding—but to highlight its limitations and demonstrate where UED offers clarity, precision, and consequence-aligned improvements in perception, communication, and technology.

1. Medium

Classical View: Ether / Electromagnetic Field

- Historically, classical theories relied on the luminiferous ether, later replaced by the electromagnetic field, which was seen as a passive backdrop.
- The medium had no intelligence—it simply carried the wave like a rope carries a pulse.

UED View: Structured Vacuum Energy (SVE)

- In UED, the medium is not passive, but structured, dynamic, and responsive.
- SVE is not "filled" with waves—it's conditioned to

align with ripples based on readiness, allowing signal realization only when specific field criteria are met.

- This explains context-sensitive communication, instantaneous alignment (like entanglement), and why space is not truly "empty."

2. Signal

Classical View: Oscillatory Transfer

- Signals are considered to be oscillating disturbances, physically moving from point A to B.
- Everything from sound to light to radio is treated as a form of repetitive vibration across time and space.

UED View: Consequence Through Alignment

- Signals are not transferred—they are realized.
- The ripple doesn't travel—it simply aligns the field between source and observer to create a consequence.
- What you "receive" is not motion—it's a resolved consequence of shared field dynamics.
- The ripple activates the field, not traverses it.

3. Speed

Classical View: Constant (c)

- Light and other EM waves are considered to move at a fixed speed (c = 299,792,458 m/s) in a vacuum.
- This speed is treated as a fundamental limit across all reference frames.

UED View: Observer- and Field-Dependent

- Speed is a derivative, not a constant.
- The Alignment Threshold Time (ATT) dictates when the signal can be resolved.
- Depending on RAC and ATT, the perceived "speed" of communication can vary.
- This explains why some signals arrive "instantly" (entanglement, perception), while others seem delayed or lost (space communication anomalies).

4. Transmission

Classical View: Emission → Propagation → Reception

- A source emits a wave, which propagates through space and is eventually received by a detector.
- The process is linear, with energy assumed to move *continuously from source to receiver.*

UED View: Field Registration & Ripple Resolution

- The source modulates the field, embedding intent or data as a ripple.
- The ripple is not in motion—it is registered in the field.
- The receiver aligns to this ripple if and only if RAC and ATT are satisfied.
- Thus, the event we call "transmission" is simply the consequence of matched readiness, not a physical journey.

5. Failure Cause

Classical View: Noise / Interference

- Failures are explained by external disturbances—random noise, overlapping waves, or signal degradation over distance.
- The focus is on external problems, often ignoring the role of the receiver's capacity.

UED View: RAC/ATT Mismatch or Misalignment

- Failure is not interference—it is misalignment.
- If RAC (Ripple Alignment Criterion) is not met, the signal does not resolve, no matter how "strong" the source is.
- If ATT (Alignment Threshold Time) is too short or too long, the ripple either doesn't register, or registers in distorted or noisy form.
- Communication isn't lost—it never aligned to begin with.

Conclusion:

Classical wave theory described what appeared to work based on observable effects. But UED dives into why communication happens—or fails—at the ripple and field level.

It reframes communication not as movement across space, but as the intelligent consequence of structured alignment. And this changes everything—from signal design to human perception to how we build future technology.

Concept	Classical Wave Theory	UED Framework
Medium	Ether (historically) or electromagnetic field as a passive backdrop	Structured Vacuum Energy (SVE): active, conditioned, and consequence-responsive
Signal	Oscillatory transfer of energy through a medium	Ripple alignment producing field-registered consequence; not actual transfer
Speed	Constant value (c) in vacuum for light and EM waves	Dependent on RAC and ATT; speed = 1/ATT; varies by observer and field state
Transmission	Emission → Propagation → Reception; treated as linear movement	Registration in the field → Observer alignment → Consequence realization
Failure Cause	Noise, interference, attenuation	RAC/ATT mismatch, unresolved ripple, or misalignment in the field

Appendix C: Suggested Experiments

As we reframe communication technology and perception through the lens of Universal Energy Dynamics (UED), practical validation becomes essential. The following conceptual experiments are designed to provoke a new class of inquiry—focused not on signal transmission as oscillation, but as consequence born from alignment within the Structured Vacuum Energy (SVE) field.

1. Ripple Delay Mapping in Variable Alignment Environments

Objective: To map how ripple arrival (perceived signal consequence) is delayed or advanced depending on the degree of field alignment and surrounding ripple clutter.

Setup:

- Use a controlled environment with multiple ripple emitters (pulsed signal sources).
- Introduce variable field conditions (electromagnetic density, thermal gradients, material boundaries).
- Employ high-precision detectors at different distances and angles.

Expected Insight: The measured delay in signal perception (not transmission) will vary depending on RAC (Ripple Alignment

Criterion)—confirming that "travel" is a perceptual illusion and alignment is the true gatekeeper of signal realization.

2. ATT Sensitivity Tests for Perception Thresholds

Objective: To determine the lower and upper thresholds of Alignment Threshold Time (ATT) in biological and artificial observers under different cognitive or hardware processing conditions.

Setup:

- Present ultra-brief visual or auditory pulses to human participants using randomized intervals (10 ms to 300 ms).
- Use EEG, reaction time software, or image flash techniques.
- Repeat with AI or sensor-based receivers (camera + AI, digital microphone arrays) with adjustable sampling rates.

Expected Insight: Biological and digital observers will exhibit sharp boundaries beyond which they fail to register ripple information, supporting the UED claim that signal perception is not continuous—but conditional.

3. SIM Identity Field-Locking Simulation

Objective: To demonstrate how a unique receiver identity (like a SIM card) functions as a field-locking mechanism that filters incoming ripple consequences.

Setup:

- Use programmable receivers with unique encryption keys (SIM emulators).
- Broadcast a shared ripple-modulated field (Wi-Fi or RF pulse with embedded data).
- Monitor which devices resolve the message based on field-locked identity.

Expected Insight: Only receivers with matching ID-locking conditions (aligned RAC & ATT + encrypted intent) will resolve the signal—proving that transmission does not guarantee realization without identity alignment.

4. Ripple Collapse Propagation in Closed vs. Open Fields

Objective: To examine how ripple-based signal collapses or sustains depending on field openness, boundaries, and anchoring systems.

Setup:

- Use ripple-generating hardware (pulsed EM sources) in two types of enclosures:
    - Open field with minimal obstruction.
    - Closed chamber with structured boundaries (Faraday cage, layered materials).
- Place synchronized receivers inside and outside both enclosures.

Expected Insight: The behavior of ripple collapse (signal fading, resolution timing, or persistence) will differ dramatically based on SVE shaping—validating UED's claim that communication is a function of field openness and anchoring, not of direct emission strength.

These experiments do not attempt to confirm old paradigms using new tools. Instead, they are designed to replace old questions with better ones.

Insight Over Inference. Alignment Over Transmission. Ripple Over Wave. This is the shift UED invites—and these experiments begin that journey.

Appendix D: Reverse-Engineered Modulation

The Hidden History

Modern modulation—AM, FM, QAM, etc.—was not born from deep theoretical insight. It was born from necessity. Early engineers noticed that varying a carrier wave's amplitude or frequency allowed more precise information delivery, especially in noisy or shared transmission environments. But what were they really doing?

In UED terms, modulation is the art of encoding **intent** onto ripple-ready carriers within a structured vacuum field. The "carrier wave" was simply a **ripple proxy**—a medium pre-aligned to resolve incoming shifts. By adjusting a ripple's **pattern, density, or phase timing**, engineers inadvertently created tailored ripple signatures that the receiver (when properly aligned) could resolve into intelligible signals.

This means:

- Modulation is not signal alteration—it's **ripple choreography**.
- Carrier waves don't carry—they **guide field resolution**.
- Demodulation isn't decoding—it's **alignment re-interpretation**.

So yes, modulation was reverse-engineered. Not as electromagnetic theory first, but as trial-and-error **alignment engineering**.

"They didn't ride the wave. They shaped the ripple."

Appendix E: Reframing Sound in UED

Beyond Longitudinal Waves

What we call "sound" is ripple-induced alignment—not a mechanical wave. In classical physics, sound is treated as a longitudinal oscillation of pressure zones through a medium like air. But UED proposes something fundamentally different:

1. **The Source Doesn't Send—It Disturbs:** A vibrating object injects ripple sequences into the medium. These aren't sinusoidal waves—they're energy-influenced alignment shifts in structured vacuum.
2. **The Medium Doesn't Transfer—It Aligns:** Air molecules don't "pass along" vibration. They ripple-register and phase-delay energy patterns in the field. No matter moves—but field tension realigns locally.
3. **You Don't Hear the Sound—You Resolve It:** The ear doesn't interpret pressure—it resolves ripple congruence. The cochlea acts as a ripple decoder, identifying patterns that resonate with neural circuits.
4. **Standing Waves = Recursive Ripple Convergence:** So-called stationary waves are simply recursive ripple alignments where source and boundary match field tension. Resonance is consequence, not causation.

Classical View	UED Perspective
Sound wave propagates in air	Ripple fields resolve through air alignment
Pressure variations cause hearing	Ripple congruence activates hearing mechanism
Speed of sound = fixed in air	Delay is field- and observer-specific (ATT-driven)

"You don't hear a wave. You resolve a memory encoded in the ripple."

This redefines not only sound, but all auditory communication as ripple-aligned interaction. No "longitudinal wave" ever traveled. Only structured consequence did.

Acknowledgments

To the engineers—those unsung alchemists of consequence—who built bridges with no blueprint, antennas with no understanding of the field they were truly shaping, and devices that worked *not because of theory*, but in spite of it. You stumbled forward, not out of ignorance, but with courageous intent. You proved, again and again, that **knowing is often slower than making**, and progress is rarely explained before it is felt.

To the thinkers who dared to pause when the world was rushing, who questioned the unquestioned, and asked not "how it works," but "what is really happening?" You made room for new frameworks when old ones grew too rigid. You reminded us that *truth doesn't scream—it resonates*.

To the independent researchers—isolated, ignored, yet persistently aligned. This book stands as a ripple born of your legacy. Not a reaction to rejection, but a response to reality. Some of you never published. Some never found an audience. Some were silenced. But the field heard you. And so did we.

To those who believe that discovery must come with credentials—may this work gently unsettle your faith. And to those who believe that all credentials must be undone—may this work remind you that alignment matters more than ideology.

To the readers who stayed—curious, skeptical, open. You didn't just follow the words. You listened to the field in between them. This book was written for you, but not only by us. It is a **consequence of everything that came before**, refined into clarity, and offered not as an answer—but an alignment.

And finally, to the **Structured Vacuum Energy**—the quiet, the constant, the carrier of all that moves without moving. Long before we invented microphones or machines, you were already listening. Every realization, every ripple, every connection... was always waiting in your stillness.

Thank you for holding the message until we were ready to hear it.

About the Author

Sandeep Chavan is an independent researcher, engineer by training, and a prolific author of over **100 books** spanning physics, cosmology, consciousness, education, technology, and the future of human intelligence. A relentless challenger of rigid paradigms, he doesn't just question mainstream science—he reconstructs it.

At the heart of his work lies **Universal Energy Dynamics (UED)**, a revolutionary framework that replaces force-based, motion-centric models with a ripple-based logic of **consequence, alignment, and field readiness**. UED challenges long-held assumptions about light, time, waves, and information—not with mysticism, but with sharp clarity rooted in observable outcomes.

Unbound by institutional filters, Sandeep works independently—yet his contributions are anything but isolated. His writing bridges **science, philosophy, and engineering** into one coherent language of reality. Whether decoding gravity through structured vacuum fields or reinterpreting modern communication as ripple alignment rather than wave transmission, his insights dismantle the unnecessary and illuminate the overlooked.

He believes that intelligence is not a matter of computation or tradition—it is the ability to align with truth when all noise is removed. Through UED, he invites readers, researchers, and future engineers to step outside the simulation of explanation and listen instead to the **silent architecture of consequence**.

Sandeep's work speaks to those who feel the limits of current models, who see the cracks in well-packaged theories, and who are willing to look deeper—not for answers, but for alignment.

Disclaimer

This book presents a conceptual and theoretical framework—Universal Energy Dynamics (UED)—that reimagines the foundations of communication, perception, and signal propagation through a ripple-based lens. The ideas expressed herein challenge prevailing scientific interpretations, such as classical wave theory and conventional electromagnetic transmission models. They are intended to stimulate intellectual inquiry, not to invalidate the dedicated work of the broader scientific community.

This work does not claim to replace any officially validated scientific doctrine, educational curriculum, or legally accepted engineering standard. It is an alternative exploration built on the author's independent research, philosophical insight, and observational reasoning.

While many of the concepts discussed may appear to contradict mainstream explanations, they are offered with the intent to expand the conversation—not to create conflict. The author respects the legacy of scientific discovery while questioning its blind spots and proposing refinements grounded in field alignment and consequence logic.

Readers are encouraged to engage critically with the content, compare frameworks, and pursue further learning through study, experimentation, and thoughtful discourse. Any application of the theories described should be undertaken with caution, context, and academic responsibility.

In essence, this book is an invitation to think beyond, not a directive to dismiss.

Don't miss out!

Visit the website below and you can sign up to receive emails whenever Sandeep Chavan publishes a new book. There's no charge and no obligation.

https://books2read.com/r/B-A-EPGPC-GFVPG

BOOKS 2 READ

Connecting independent readers to independent writers.

Did you love *It was Never a Wave Signal*? Then you should read *Frequency*[1] by Sandeep Chavan!

[2]

What if light was never sent to you—only waiting to be realized?What if frequency isn't carried by waves, but extracted by observers?What if science never discovered truth, but only tuned into what it was ready to receive?

In *Frequency: The Compass of Consciousness and the Illusion of Light*, independent researcher and philosopher **Sandeep J. Chavan** presents a groundbreaking reinterpretation of modern physics and perception through the lens of **Universal Energy**

1. https://books2read.com/u/bODMWg
2. https://books2read.com/u/bODMWg

Dynamics (UED)—a model that places resonance, intent, and structure at the heart of how reality is realized.

This is not a book about equations. It is a book about **alignment.**It challenges our inherited assumptions about photons, frequency, observation, and even the nature of scientific discovery itself.From rethinking the photoelectric effect to questioning the objectivity of $E = h\nu$, Sandeep guides the reader through a paradigm where:

Light is not emitted—it is **realized through resonance.**Frequency is not inherent—it is **observer-dependent.**Observation is not passive—it is **intent in action.**Science is not final—it is **structured tuning** into the field.

Blending physics, metaphysics, consciousness studies, and system design, *Frequency* unfolds as both a scientific discourse and a philosophical compass. It invites the reader—not merely to understand the universe—but to realize that **they are already a structure within it**, shaping what can be seen, known, and made real.

Featuring original laws (such as the **Law of Consequences**, the **Half-Truth Principle**, and **Chavan's Law of Frequency**), detailed diagrams, reader-friendly glossaries, and rich appendices, this book is a complete intellectual toolkit for those who believe science has more to say—and more to remember.

If you've ever suspected that reality is more relational than mechanical, more participatory than objective, and more conscious than random—this book is your tuning fork.

You do not find the truth. You align with it.

Read more at https://www.sandeepjchavan.com/.

Also by Sandeep Chavan

Chavan's UED research

CMGED Research Report

Chavan's Cosmic Continuum (C^3)

Frequency

Ripples of Reality

Reality Realigned

The Great Realignment

I – The Cosmic Leverage

Chavan's Cosmic Theory of Relativity (CTR)

Chavan's Theory of Light

Feeling Like a Physicist

TIME – The New Definition

$E \neq mc^2$: The Collapse of Classical Assumptions

The Chavan Codex

Chavan's Atomic Model

Echoes of War: The Russia-Ukraine Conflict and Its Global Ripples

Shattered Lives

The War Economy
Borders Redrawn
Arms and Allies
Collateral Damage
Justice in the Ashes

Somewhere in the Darkness of the Cosmos
The Blueprint of the Universe
The Cosmic Breath
From the Edge of the Universe

The Eternal Sage Chronicles
The Sage and the Sovereign
The Family Pillar
The People's Flame
The Warrior's Soul
The Bridging Mentor
The Guiding Light

The Exodus
The Last Cycle: 2075

The Fundamental Theory of Life & Existence
Energy in Action: The Soul Demystified

The Active Mind
Energy in Connection
Death: The Disconnection
Subconsciousness: The Chosen Disconnection
Desires: The Engine of Life

The Restoration of Human Knowledge
Restoring Bhagavad Gita

The War of Thoughts
Cracks in the Known
The Human Game
The Divine Dilemma
The Original Intent
The Answers

Ultra Modern Life Philosophy
The Duality of Laughter and Tears
Take It Easy
Compare to Accept

Standalone
It's Not AI, It's AHI - Amplified Human Intelligence
The Decision Paradox

The Four Sapiens

Malicious Script of Indian Polity

The IIT Legacy & Global Impact

The IIT Legacy & Global Impact

You are just a Version of Your Original

Win the Game You Didn't Choose

Always Keep Your Bags Packed

Reveal to Shield - Learn the Game of Facades

Engage Beyond Elections

Culture, Identity & Change: The Evolution of Indian Society

The Triangular Dynamics

Never Trust Without Doubt

If You Can't Change the Country, Leave the Country

Live Fearless with Fear

India: The Investment Magnet

Realignment

Powering a Billion Dreams

Universal Skill

Extension: From Brush to Algorithm

Human Side of Einstein

The Future Knows No Lines

Let's Bring the Generations to the Table

Unified Energy Dynamics

A Beat That Never Dies

Strategic Horizons

Right From Scratch

The Forgotten Innovators

Physics Fixing

The Savior Illusion

The Selfish Side

The Age of the Weakest Sapiens

Neural Wars
The Absorption: There Is No Death
Strange Therapist
Bubbles: Become a Drop
Beyond the Blue
Half-Scholar
The Law of Truth
Failure
Philosophy 2.0: A Structural Manifesto
Chaos: A Gentle Guide to the Unresolvable
Self-Help: New Syllabus
Rippling
Game: How to Play the Share Market
Out of Syllabus
Saffronized: The Code Behind the Crown
The Anger
Moksha: The Liberation
Chavan vs Einstein: The Universe Was Never Curved
Rocket Science: The Ripple Gambling
The Cosmos: The Still Universe
The Taste of Silence
Chavan's Post-Materialist Human Model
10-Second Rule
Chavan vs Hawking
It was Never a Wave Signal

Watch for more at https://www.sandeepjchavan.com/.

www.ingramcontent.com/pod-product-compliance
Lightning Source LLC
LaVergne TN
LVHW090359210926
848480LV00002B/396
* 9 7 9 8 2 3 1 8 7 8 7 8 9 *